First published 2021

Published under licence by Brown Dog Books and
The Self-Publishing Partnership, 7 Green Park Station, Bath BA1 1JB

www.selfpublishingpartnership.co.uk

ISBN printed book: 978-1-83952-215-4

Cover design by Lexi L'Esteve

Printed and bound in the UK

This book is printed on FSC certified paper

FSC www.fsc.org MIX Paper from responsible sources FSC® C013604

The Pointing Dog

its selection, rearing, training and work

by

Steve Wright

BROWN DOG BOOKS

Contents

DEDICATION

This book is for the many enthusiasts who have embarked on the treacherous waters of hunt/point/retrievers (HPR) ownership and training. It is a great voyage of discovery, and one which involves frustration, stamina, good humour and patience in equal measure. Many of you have unstintingly shared your experiences and hard-won knowledge, and this willingness to encourage and support new owners is one of the hallmarks of the sport. I have been a grateful recipient of so much help from you, and now it is my turn to give something back.

Mostly this book is dedicated to those wonderful (and infuriating) workaholic dogs that have been my companions in the field over so many years. They have given so much affection and sheer hard work, have continually confounded me – with their absolute brilliance and exceptional noses, and deserved every ounce of the love and admiration I have for them.

ACKNOWLEDGEMENTS

The reason for writing this book was to encapsulate, in one document, the many letters, emails and telephone conversations I have had with HPR owners over the years – requesting advice, help, support and counselling! Very often I have been asked to suggest a book that can be referred to any time. I never found one, when starting out with my first pointing dog many years ago. What a boon it would have been.

Therefore, my first acknowledgement must be to those of you who asked the questions that have provided the 'skeleton' of this book's content. My own knowledge came firstly from my parents, then my many colleagues and friends in the working dog community. Much of it since has been learned by experience, and the opportunity to discuss this with fellow enthusiasts. Many thanks to Maurice Cook and Norma Ansell, Ray Davies, Fiona Wensley, Lloyd Birch, Dave Anderton and all the others!

On the basis that one picture is worth a thousand words I have relied heavily on my own amateurish stock of photos, for which I apologise, and many more from much better photographers. Most are HPR owners and enthusiasts, such as Sue Axtell, Christine Carpenter, Andy Coveney, David Hudson, Jane Lowe, Carolyn Perks, Selina Sleany, Susan Stone, Debby Wearing and others too numerous to mention.

Lastly my thanks go to my partner Sue Connolly, for unstinting support, volunteering to be an unpaid demonstrator, for supplying endless cups of coffee and for putting up with my obsession with dogs.

Foreword

My life as a dog trainer began with training our pet dogs for obedience, which required them to display no initiative. On starting work, I trained guide dogs for the blind for over twenty years, and these wonderful dogs need great steadiness and lots of initiative to work well. Many of my colleagues followed a variety of dog-based sports, such as sled-dog racing, gundog work, agility, tracking and police work. As a contrast I took up whippet racing. These brilliant canine athletes had to be completely unsteady, and were simply encouraged to total commitment to the chase. They had no need of decision-making, and I loved the superb physical specialisation that had created a dog that could reach 40 mph in just a couple of seconds, and could duplicate this effort repeatedly.

In 1987 I got my first HPR (hunt/point/retriever) after I had eventually turned my falconry hobby into a business, and became fascinated by the complexity of training and working them. They are supreme examples of dogs bred specifically to fulfil a very complex series of tasks – both mental and physical. They have phenomenal drive and initiative, but need to remain obedient – even at a considerable distance from the handler. This is only possible because of their contradictory qualities. They are dogs that can run for hours, yet freeze in one stride's length and remain motionless for several minutes. Which are sensitive enough to delicately carry live birds, while being sufficiently tough and bull-headed to face the thickest bramble cover. Such dogs need to be bred properly as well as receiving the best of training – and they must be worked. HPRs are not designed to be fireside rugs!

Dog breeding has always been a passion for me. I have been fortunate to have produced some outstanding dogs in different spheres, but finding them the right home is vital to their success. Anyone who buys a working-bred dog gets a complete package, that predetermines what that animal wants to do. All the HPR breeds are relatively new to the British dog world, and their whole style of working is at odds with how we traditionally use gun dogs. Even many good gun-dog trainers find they are a 'dog too far'. Consequently, there are few experienced people who can give the new HPR owner advice on how best to use and satisfy their pet.

As a breeder, I saw that I had a responsibility to provide backup to the buyers of my pups. This extended to owners of pups of other HPR breeds, and from other breeders. Through the Brittany Club of Great Britain I helped implement a mentoring scheme of local, more experienced owners and have run breed-specific training days (sometimes attended by other HPR breeds which did not have this local support). At my local gun-dog society, I am part of a team of trainers offering specialist instruction for owners of all the pointing breeds. This book has evolved from the variety of different 'handouts for common problems' that I have accumulated through years of correspondence with novice handlers. My belief is that there are too few good books on this specialist subject.

Organising its layout and content has been quite difficult, because pointing dogs develop a whole 'portfolio' of varying, and possibly contradictory, skills. It made sense to describe, in separate chapters, the development of each individual element of their work, but actually, you will be teaching their complete spectrum of skills simultaneously. Progress in the different areas will be readily, or less easily, achieved at the same stage in a pup's development. There is no simple progression, and you cannot just wait until you have finished all the exercises on retrieving, before moving on to pointing. Start at chapter one and work steadily through to the end. Your timetable will be dictated by the temperament of your individual dog and your own opportunities. For example, a naturally biddable dog will be more inclined to obey readily than the youngster with a precocious hunting drive.

Quite possibly you will find yourself continually referring to previous chapters or looking ahead, with the aid of the index, to see if your current problem affects other aspects of the training. Ideally you will read through most of it first, so that you gain an overview of the holistic approach that HPR training demands. My hope is that this will become a 'well-thumbed' occupant of your bookshelf.

I must apologise to owners of other HPR breeds if this book seems to concentrate on the Brittany. Largely this is because this has been my personal experience, and it was easier for me to get photos of them to illustrate the text. However, all HPR breeds have the same skills to learn and the same basic qualities to work with – but the Brittany is the most headstrong and enthusiastic of them all! I promise you that if it works for a Brittany it will work for your breed!

INTRODUCTION

From their original wolf ancestry man has succeeded in breeding dogs in a bewildering variety of sizes, shapes, colours and coats. He has also selected, and managed to fix, behavioural patterns so that predicted, inherited traits will evince themselves in the resulting generations. This has been achieved by choosing breeding animals for their individual personalities. It could only have been managed by prolonged study of them and a recognition of those varying traits that man thought were desirable.

How we arrived at the pack behaviour of scent hounds is easy enough to understand. It is no different from wolf behaviour, but we have changed the coloration to aid easy visibility, except in the Otterhound. For this specialist breed visibility is immaterial as the pack works within the tightly defined banks of the river. Where hounds are intended to range over wider landscapes at greater speed, man has increased the use of the hounds' voice to help him follow the chase. By only breeding from those individuals who showed the greatest willingness to use their noses, we have achieved breeds that rely on scent almost exclusively. Selection of more placid temperaments has also meant that the pack is easier to house together in numbers, and can be more readily controlled. Size is determined by the intended quarry, and the intention of the huntsman to follow on foot or horseback.

The gazehound is another hunting type, but it is happy to work in isolation and relies on out-and-out speed. In the wolf pack the front runner has more than just fleetness of foot. Because it is able to overtake its quarry by speed it has learned to ignore scent (which would slow it up) and relies instead on its eyesight. It also displays a willingness to leave the pack behind, which has led to a reduction in pack behaviour, and has become more reliant on its master for companionship. It not only identifies with humans but it is prepared to bring its quarry back to us, in a corruption of the behaviour that would see it return to the pack with food. A similar selection process has then seen the development of retrieving breeds, which have a compulsion to fetch all sorts of things for our approval and reward.

We have harnessed the wolf's natural territorial drive, and perhaps used dogs with particularly strong maternal instincts, to produce dogs to guard our livestock, property and possessions.

When a wolf pack is hunting there are always individuals, perhaps less courageous or just slower than their fellows, who show a tendency to flank the quarry rather than to tail-chase it. This makes it easier for the more aggressive pursuers to catch it, as the quarry is less able to jink and turn. These flankers became the ancestors of the herding breeds, where we exploited a negative trait and developed it into a specific skill. Another negative trait is that of stalking. Instead of an outright attack, the slower (or should it be more timid?) wolf learned to use a stealthy approach for fear of startling its quarry. Sometimes it stops dead rather than give the game away.

This pattern of behaviour is the basis of pointing or setting. In pointing dogs it has been developed to the extreme so that the dog stops, because its nose tells it there is quarry there, without needing to make visual contact first. Moreover, as long as the quarry keeps still, the dog seems unable to move, and the two creatures remain locked in a frozen and deadly stand-off. As long as the quarry remains motionless the dog will wait, sure that something is there. And as long as the dog waits the quarry remains immobile, rather than reveal its whereabouts.

It will be seen that pointing is something of an anxious reaction. The dog stops initially because it is uncertain. Very dominant and assertive individuals, although bred for pointing, may not point at all or may be unsteady and unreliable on point. Stopping through uncertainty is alien to their temperament. Possibly, by being more dominant ourselves, we could undermine their assumed status and reduce their confidence, thus inducing better pointing behaviour.

This sixteen-week Brittany puppy is already showing hesitancy when coming across strange scents

In contrast some dogs are naturally so sensitive that they live in a state of continual uncertainty. Their work is frequently marred by continual 'false' points, where they stop for every little scent that might indicate game. These dogs may also stop so far off game, at times, that they are useless as indicators of the presence of quarry. It may be worth letting such dogs get a bit out of hand, allowing them to become less submissive. Future control would be easily re-established, as they are basically subservient, and will soon accept our right to impose our will on them.

It is our good fortune that the wolf, or its domestic descendant, is happy to include us as members of its pack. Its understanding of social hierarchy fits comfortably with our own, enabling us to achieve a ranking and status that it will recognise. However, it would be wrong to assume that we are automatically going to be at the top of this mutually acknowledged 'pecking order'. In pet dogs the majority of behavioural problems arise from the animal's unwillingness to accept our authority, despite our expectation that this will automatically be the case. Only by virtue of the fact that we have selectively bred dogs that are easy to control, do most people manage to live in peace with their pet. Very often it is apparent that the dog rules the household, with an iron fist in the traditional velvet glove. It is the owners who are avoiding confrontation.

Anyone seeking to work a dog must recognise the original purpose for which each particular breed was developed. Within that breed some individuals will be better or worse at it than their fellows. Often a good dog is successful largely because it has been trained by someone who suits it temperamentally, and vice versa. In other hands it might not have realised its full potential. Selecting a new puppy is not only a case of identifying the breed which is best suited to the intended task, but also finding the kennel or strain that has a typical temperament that will be compatible to yours. Even then each puppy will vary within a litter, and a degree of self-knowledge may be needed to make the right choice. Are you better at building up a sensitive dog's confidence, or are you more comfortable harnessing the drive and enthusiasm of a more boisterous pup?

Following selection there is the vital task of rearing your puppy. I am not talking about nutrition, although that is clearly important, but the way in which you handle it from the start. Some people always have nice dogs. This is not an accident but rather that they automatically expect the dog to conform to their ideas of good behaviour. Being for the most part obliging souls, the young puppy learns to fit in. Do not underestimate the importance of this early period in your puppy's development. Lessons learned at this stage will be imprinted on the dog's mind and will stay with it forever. Much of this is done through play, as would be the lessons learned in nature. By contrast the over-zealous trainer, who consciously teaches the pup set exercises, will make great strides at first, but may suddenly find the dog has had its enthusiasm and initiative knocked out of it.

Like any other craftsman the dog trainer must invest time in preparing his materials properly. You may throw a beautiful pot on a wheel, but if the clay has not been thoroughly kneaded and 'wedged' to remove all the air bubbles, it will shatter in the kiln. Similarly, a cabinet made from green wood will split and warp, no matter how straight the wood was cut and how accurately the joints were formed. For the dog trainer the pup's formative months are just as vital, if it is to be the finished article that was intended.

Training working dogs involves creating the ideal balance between initiative and obedience. In most situations we can tell what our dogs are doing but, with anything relating to scent, the dog exists in a world of which we are ignorant. This means we are unable to check on it and can only assume that it is doing the right thing – which can test the nerve of novice handlers to the full. It can be very tempting to second-guess your dog but experience will tell you that it is actually working, despite giving every appearance of rioting. The only way you can be sure that it is behaving is to have such a degree of control over it, at all other times, that you know it would not dare to mess around.

Why pointing and setting dogs were developed?

In the modern age we have some very powerful guns. Their shot can out-speed any animal or bird but they have a surprisingly limited effective range. The modern 12-bore shotgun is of little use over more than 35 metres. Therefore, the sportsman needs to get close to the intended quarry. Fashionably this is done by driven game, where the gun stays still and teams of beaters drive the game towards the guns. Faced with a cloud of birds streaming towards them success is almost guaranteed. Some expert shots are able to take very large bags of game in this way.

Large-scale driven shoots depend on stocked and reared game being present in numbers, and a whole industry has grown up around this. Landowners, game farms, keepers, beaters, sporting and corporate agents, gunsmiths, loaders, dog trainers and handlers, game dealers, gentlemen's outfitters, even car manufacturers all derive income from this industry. It is a major employer in the countryside and it all depends on the shotgun.

Going back a century or two we had only muzzle-loading guns and flintlocks that were slow and inaccurate. The shooter's ability to get several shots off in quick succession was much more limited – so a barrage of birds was of little use. What was needed was to know in advance exactly where the quarry was, and to have ample time to prepare. The modern spaniel, flushing game at random, was OK but had to work very close if it was not to put birds up out of range of its owner's gun. Where game was scarce or scattered the spaniel was not sufficiently productive, as it did not cover enough terrain.

Happily, there were dogs available from an earlier age that served this purpose well. These were dogs used to net small songbirds, such as larks, by pointing where the birds crouched so that their owners could throw a clap net over them. Falconers too had always needed dogs that would range widely to find scattered game, yet would not flush them. Such dogs could scour vast tracts of land and would freeze once they had located quarry. This enabled the falconer to position the hawk or falcon to best advantage, allowing for wind direction and contours, before flushing the game. In a successful hunt the pointer or setter was not required to retrieve – a more modern skill the shooter valued.

To remedy this omission the early sporting gun, with his 'fowling piece', had two options. Sometimes a second dog could be used, purely as a retriever, leaving the pointing dog to continue its quartering of the landscape in search of even more game. Alternatively, the shooter might find individual dogs that would bring birds back. Intentionally or otherwise, the resulting cross between pointers and retrievers sometimes incorporated both these skills. Thus, the HPR (hunt/point/retriever) was born.

Today the pointers, setters and HPRs, as a group, do not fit in with the modern obsession for large bags of driven game. Their use is mainly confined to 'walking-up' days, rough shooting and falconry. Very rarely do modern guns shoot over pointers, which is a great shame. It is a style of sport that involves the participant more, requires greater fieldcraft and has less impact on game numbers. I wonder if the economics of running six 'walking-up' days for four or five guns, with an average bag of thirty or so birds per day, would be more viable than a dozen guns enjoying one driven day with a bag of three hundred. The savings in beaters' wages alone would be considerable.

New uses are being found for this group of dogs. They are used for counting grouse and partridge on some shoots, and for gathering up the hen pheasants at the end of the season. The first time I saw this done was by an elderly gentleman, with his equally venerable English Setter. He simply carried a lady's three-pronged garden fork with him, having first removed the central tine. When the dog came on point he would carefully pin the birds to the ground, with their wings held against their bodies by the remaining outside tines. He then gently picked them up by hand and put them in a sack.

The scenting capabilities and persistence of the pointers and HPRs are also being made use of by our security forces in an increasing diversity of fields. They are particularly useful for outdoor searches as they can cover so much ground. How easily the instructors will find adapting to the demands of training the HPRs will be an interesting test of their imagination and flexibility.

In this book I have set down what I have learned from over thirty years of working HPRs. Before this I trained many different breeds of dogs in varying disciplines, and have successfully bred good working dogs in some very diverse spheres. More experienced trainers than me will, no doubt, be able to pick faults in my methods, and I do not dispute that their techniques are effective for them. Wherever dog folk gather there will always be a lively debate, and this is one of the charms of the sport. I hope you will find this book to be both an accessible and logical guide that will help you to achieve a happy and rewarding working relationship with your HPR.

Chapter 1
The Pointing Breeds

The origins

The origins of pointing dogs go back a long way and it is a brave person who can make claim to definitive knowledge on the subject. My research has been extensive, but not exhaustive, and the following comments should be seen for what they are – an informed guess at best. While each breed club usually offers a history of their particular breed, which in isolation appears clear and comprehensible, the overall picture is much more confusing and often contradictory.

The different breeds of dogs that we have today were mostly 'fixed' by enthusiasts in the late nineteenth century. Before this time dogs were usually thought of as being 'of a type' – governed by their working role rather than their appearance. Hence, we recognised terriers, spaniels, gazehounds, herding dogs etc. Only after Gregor Mendel, the Augustinian monk who became the 'father of modern genetics', discovered the rules of genetic inheritance, did the concept of deliberately breeding for appearance and performance become fashionable. Early standards were based on outstanding individual dogs of the time. The history of breeds since then has been easily traced. Crossing two 'types' to introduce individual traits was usually unsuccessful, such as the foolhardy introduction of foxhound blood to pointers in an attempt to increase the latter's size, strength and visibility. Unfortunately, the overwhelming quality of the scent hound is to chase, which immediately detracted from the pointer's whole reason for existence. By contrast, the crossing of dogs of a similar type was found to be more productive, such as the introduction of German Wire-haired Pointers to Hungarian Vizslas to improve the hardiness of the latter without reducing its pointing ability. Before this the waters are a good deal murkier.

Across Europe there existed dogs that were used to point game for falconers or to aid the netting of birds. These came in a variety of sizes. Largest were the *braques*, which seem to have been large powerful dogs with a suspicion of mastiff in their ancestry. Mentioned as long ago as the third or fourth century BC, the *braques* must have been the forebears of the Bracco Italiano and the Italian Spinone. The former is mooted to have arisen from a blend of the *braque* and the St Hubert Hound from which comes the Bracco's head and typically heavy ear carriage. The Spinone, on the other hand, although not dissimilar in size and build is suggested to have got its coarse coat from the *Seguigo Italiano*, or from *coarse-haired setters* that were brought in from the Adriatic.

Possibly the greatest influence on the breeding of pointing dogs was the *Spanish Pointer*. A painting by George Stubbs depicts a powerful, muscular dog of a heavier type than the modern English Pointer. Given a dash of spaniel blood the *Spanish Pointer* gave rise to the various setters, the English Pointer, and seems to have contributed considerably to the various German breeds. However, the German Long-haired Pointer is claimed to get its coat, and possibly its more placid temperament, from the *Spanish Fowling dog*, which may have been either a distinct or a related breed. The German Wire-haired Pointer may have some *griffon* blood, rather than getting its coat from the same source as the Italian Spinone. There is also mention of a *Yellow Turkish Dog* that may have influenced the Hungarian Vizsla's colour at least, and possibly that of the Weimaraner also.

Finally, the French *fougere* was said to be a small 'spaniel-type' dog of black and white, or liver and white colouring. It is believed that these were bred to imported pointers and setters to create the Brittany and the Small Münsterländer, conspicuously the smallest breeds of the group. Certainly the coat and colours to be found in this breed would support this. The drawing '*Hawking for Pheasants*' by Francis Barlow, dated 1684, shows dogs of a type that would be recognisable today.

THE POINTERS AND SETTERS

These are all tall dogs (55–70 cm at the shoulder) and have been bred to work large tracts of open country such as the grouse moors. Of the traditional British breeds only one, the Pointer, is smooth-coated. No doubt, with the current vogue for scouring the far corners of the earth for new, rare breeds, there will be others discovered, such as the smaller (but more powerfully built), sandy coloured Portuguese Pointer, and the slightly built Braque d'Auvergne. All the setters are long-coated and the breeds are only separated by colour, build and athleticism. There will be differences in temperament and performance but, with so few being worked consistently, it may be difficult to identify strains that are bred for performance. They are not expected to retrieve although this can and does occur.

The English Pointer

This is an active, boldly marked dog with a very short coat. It comes in lemon, black, orange and liver, but usually with a substantial amount of white, although self-coloured dogs and tricolours are acceptable even if rarely seen. There is usually some ticking to the coat, but heavily roaned animals are not desired as this reduces their visibility in the field. They are tall (63–69 cms for dogs [25–27 ins] and 61–66 cms for bitches [24–26 ins]). The body is quite narrow, with a deep chest, a relatively short back and a longish, straight tail that is carried level with the back. The shoulders should be well laid back giving a smooth, powerful action with plenty of drive. In movement the front leg action extends well forward and quite high, to enable them to get through deep heather effectively. It should not be 'hackneyed' as this would inhibit the length and freedom of the stride. Pleasingly they are one of the few breeds that are not shown in a fleshy condition. The overall impression is of a strong, active, well-muscled dog that is neither coarse nor weedy.

When working, the pointer should exhibit plenty of drive and dash. As is typical among air-scenting breeds the head carriage, when quartering, is level with the back or higher. It should range over a substantial acreage, and an outrun of 200 yards either side would not be in any way excessive. Very obviously this means that a pointer can easily get out of sight of its handler. Therefore, it must be very staunch to the point (i.e. hold its point steadily) and be easily noticed. This is where white markings are very useful, as often only the top half of the dog may be visible in deep cover.

The stance, on point, will normally exhibit a level top line, i.e. the head, neck, back and tail should all be in one continuous flat line. The dog leans forward to the point of balance and it will lift a leg in anticipation of the next step. Which leg varies according to how abruptly it stops, and the immediate terrain. However, a dog that 'works on' to a scent before coming to a halt, will often lift the same leg each time. This may change according to the quarry species so the handler can reasonably predict whether it has grouse, snipe or whatever in front of it.

Photo by David Hudson

A group of pointers grouse counting. The more traditional black & white and orange & white are joined by an unusual self-coloured black dog.

On grouse moors it is usual to only work pointers for half an hour or so at a time, before resting them and releasing another. Whether this is in case they get tired, or because they become wayward if not brought back under regular restraint is unclear. Since the breed standard includes endurance as one of the defining characteristics, it is probably the latter. This may mean that one pointer is insufficient for the serious sportsman or woman and you may also need to commandeer the services of a 'dog boy' of some sort.

Pointers are definitively bird dogs. They excel on very open areas such as grouse moors and the stubble fields of East Anglia. They do not face cover well, nor enter water readily, and they have no taste for retrieving. Being so long-striding they do not cope well with rocky ground as their action can easily become disunited. Because of their growth rate it is unwise to work a pointer seriously until it is eighteen months old. They do remain active for a long time so you could easily be working one until it is getting on for ten years old.

Common injuries are wrist and pastern sprains and you might expect that arthritis would be a problem for the older dog. The fine coat means that they feel the cold, so they need to be 'rugged up' in cold, wet weather. It also means that you or your vet need to be handy with a needle and thread, given the proliferation of barbed wire in even the most remote countryside.

As a breed the pointer is usually, but not necessarily, a charming personality and benefits from human company. (However, I was once badly bitten by a pointer whose home I was visiting – it was in the garden and just came straight up at my throat without any warning!) Like all pointing dogs they are potentially neurotic and dislike being left. It is certainly useful to have them content to stay in their kennel, but they can be quite destructive even then. The current trend of having indoor, wire cages is probably a real boon for them.

Visit www.englishpointer.co.uk

THE SETTERS

The four recognised breeds of setters all fulfil the same basic working role. They are designed to cover large areas of rough, poorly stocked country in pursuit, exclusively, of feathered game. Once birds are found their task is to freeze or 'set' themselves, until their owner arrives and they are told to flush the game. Needing to cope with working deep heather on rocky moors, or boggy reed beds, the setters are all active, free-moving dogs. While the purist will identify the distinguishing points of each breed, they have great similarities for the lay person. All are between 61–69 cm [23–27 ins] at the shoulder and have silky coats of spaniel length (except where this has been exaggerated for show purposes). When working, it is fair to say that selective deafness is a prevailing trait, coupled with formidable stamina and plenty of drive.

The Gordon Setter is the largest, and is slightly heavier built. It should be black and tan in colour although small white patches on the chest are still evident in some specimens. Originally, they often had white patches, and could even be black and white without any tan. The name derives from an early patron of the breed, the Duke of Gordon, who was largely responsible for fixing the coat colour as we know it.

Photo by David Hudson

Very few working Gordons exist nowadays but there is no reason why they would not acquit themselves well should the opportunity arise. Getting one from a genuine working kennel may prove quite a challenge. I have seen a Gordon doing water retrieves, so they are potentially a pretty versatile breed. Gordons did have a history of hip dysplasia which caused extreme concern within the breed. Sustained efforts to control this have been largely successful but do make sure both parents have a better than average hip score before buying a puppy.

Visit www.britishgordonsetterclub.org.uk

The English Setter is slightly smaller than the Gordon, and a little more solidly built. The basic coat colour is white with patches of blue, orange, lemon, liver or tricolour (blue and orange or liver and orange with white patches). In all colours the white is heavily roaned and is called Belton, e.g. blue Belton or lemon Belton after one of the original kennels of the breed.

Working English Setters are sometimes quite small and whippety, but this is probably the product of systemic line-breeding, and is not a prerequisite of the working ability. I saw some Llewellyn setters in the 1960s (Colonel Llewellyn had an illustrious kennel in the first half of the twentieth century, and these were still owned by his retired gamekeeper) and they were very small and had exaggerated shaping to their skulls. Conversely, the first English Setters I ever saw working were conspicuously large, strong and had quite broad, coarse heads compared with show stock.

photo by David Hudson

Of all the setter breeds the English is probably the most frequently used in the field but they are still a rarity.

Visit www.englishsetterassociation.co.uk

The Red or Irish Setter is numerically the most common of the setters, but this is due to its popularity in the show ring and as fashionable pets. They are more lightly built and narrower than the Gordon or English. They should have a uniform deep red coat. White markings are deemed a fault but often small patches can be seen on the chest. As expected, they are very active, and it is difficult to keep weight on young Irish Setters.

Photo courtesy of David Hudson

Quite a few Irish are worked in their home country. Traditionally they are used on snipe and woodcock but they will point any feathered game. Perhaps their temperament reflects the nature of their countrymen, as they are a little anarchic and have scant respect for authority. However, with a few years of experience under their belts they can become steady reliable workers.

Anyone thinking of training an Irish should really look for stock in Ireland, and first see the parents working in the field. As a breed they have had more than their share of hereditary health problems (notably congenital night blindness or Progressive Retinal Atrophy) but this is now well-controlled by responsible breeding.

Visit www.isae.co.uk

The Irish Red and White Setter is the forerunner of the Red Setter and has been rescued from oblivion by a few devotees. It is a little more substantial than its younger relative and is easily identified by the bold red and white markings.

Temperamentally it is steadier and less volatile. There are some very well-motivated breeders of Red and Whites who are keen to develop the working side of the breed.
Anyone getting one, with this in mind, will receive plenty of help and encouragement.

Ch Charnborough Glamour girl
Photo courtesy of Susan Stone

Visit www.irishredandwhitesetterclub.com

CAVEAT: Before embarking on selecting and training any setter or pointer for work you would be well advised to see some in the field. It might make you abandon some of your preconceptions. A report of a Scottish Setter and Pointer Championships, in the Scottish Field, reported that '*the first runner, an Irish Setter, on its way from the car park to the start, distinguished itself by not only locating a covey of grouse but in catching one in the air!*' The remainder of the article suggested that some of its competitors were barely any steadier. Very obviously there are some expert setter and pointer trainers about, but also many enthusiasts who have yet to triumph over their dog's wilder side. Make sure you locate one of the former and get his or her help with your dog's education.

THE HPRS (HUNT/POINT/RETRIEVERS)

While the pointers and setters are excellent game finders and pointers they are not generally expected to bring game back to hand. Neither do they normally face cover well, and they may range more widely than is required in enclosed country. In consequence they are unsuited to much of lowland Britain, where sportsmen and women have increasingly begun to appreciate the abilities of the HPR group. It seems strange that a nation with a history of producing sporting dog breeds should never have developed its own native variety. Instead we have imported breeds from mainland Europe where walked-up hunting and rough-shooting is the norm. Even today our obsession with driven game militates against the use of these fascinating and multi-talented breeds.

Docking

Some HPRs, but not all, are traditionally docked to prevent tail damage when working in thick cover. This is still permitted in England and Wales (under the Animal Welfare Act 2006) where puppies are intended for working purposes. Scotland originally banned all docking but has since restored it for working-bred litters. The situation in other countries is unpredictable. The animal activists, with no knowledge of the damage an undocked dog can sustain in the field, continue to occupy the moral high ground and there is no doubt more countries will ban docking.

Docking can only be done by a veterinary surgeon when the puppy is under four days old. Performed with a sharp and sterile pair of surgical scissors it causes only momentary discomfort, and any bleeding is quickly staunched. Less frequently rubber bands of the sort used on lambs' tails may be used. This method eliminates bleeding and causes very little discomfort over a period of just a few days. The best docking technique I have ever seen was done by an old vet with a pair of wire cutters. The action of these compressed the edges of the dock tightly together rather than cutting the tail. He held the wire cutters closed and easily pulled the unwanted length away. There was no blood and the neatest sealed end you could wish for.

Your veterinary surgeon will need proof that the breeder intends the dog to be used for bona fide working purposes, and you will be asked to sign a form and to produce a shotgun certificate or other evidence. Once docked the puppy must be microchipped at the age of six weeks by the same veterinary practice, so that its legal docking can be proven.

The German Pointers – Wire-haired, Short-haired and Long-haired

The German Short-haired Pointer (GSP) on the left, and the German Wire-haired Pointers (GWP) opposite, are visually two coat variants of the same breed. Both can be liver and white, or black and white, with very heavy roaning; individuals of both breeds can be self-coloured. There is often a considerable size difference between the sexes and the males need firm and consistent, regimented discipline.

The vast majority of dogs winning at HPR Trials are from these breeds. They are truly dual purpose, but it

Photo by Jane Lowe

would make sense to source pups from recognised working kennels. Temperamentally the GWP has a tougher nature, having been developed for a variety of roles beyond basic hunting. This included tackling vermin such as fox and it is no surprise that some GWPs (particularly the males) can be quite aggressive with other dogs.

The GWP has contributed to other HPR breeds, notably the **Wire-haired Vizsla**, and is present in the mix which created the **Slovakian Rough-haired Pointer** (previously known as the Rough-haired Weimaraner). It is also very similar to the **Korthals Griffon,** which is indistinguishable to all but breed fanatics.

Visit www.korthalsgriffonclub.co.uk and www.slovakianroughhairedpointerclub.org.uk

The **German Long-haired Pointer** (GLP) is quite distinct in its physical features and temperament from the GSP and GWP. It is usually left undocked and in consequence has more of the look of a setter about it. It is less square in outline and appears to be a little shorter on the leg (almost like a cross between a setter and an English Springer Spaniel). The *'langhaar'*, as it is known in Germany, is only found in brown and white and solid brown. The few that appear at trials acquit themselves well and I would expect this breed to become more and more popular in the field.

All three breeds are good workers in the right hands, and can give a really stylish performance. From my own observations I feel that they can sometimes be a little jealous of their game and can 'peg' quarry rather than flush it for you. Breed enthusiasts will disagree with me.

There are lots of good examples of all three breeds being worked (and shown, as they are genuinely dual-purpose) and it should be easy to locate suitable puppies. Faults to avoid are aggression, hard mouths, pegging and chasing game.

Visit www.gwp.co.uk, www.gsp.org.uk and www.germanlonghairedpointer.org.uk

Large Münsterländer is another product of the German passion for identifying localised varieties as separate breeds. It was considered as a German Long-haired Pointer until the 1920s, when the GLP breeders refused to recognise the black and white colouring of the Münsterländer. This was claimed to arise from the introduction of Pointer and English Setter blood. More probably it was the result of jealousy and ill-feeling between supporters of the different types.

photo by Sue Axtell

Visit www.largemunsterlanderclub.co.uk

There is also a **Small Münsterländer** which resembles a Brittany in size but is not known in the UK.

Weimaraners (pronounced Vimararners) – and **Long-haired Weimaraners**

Developed in the Weimar Republic, this breed is sometimes known as 'the grey ghost'. It is a most unusually coloured dog having a very short smooth coat which is uniformly silver-grey. This description does it a disservice, as there may be tones of mouse or roe grey about it. The eyes are yellow. The long hair comes from a recessive gene and the two coat types compete equally at shows and trials.

Photo courtesy of Shirley Cooper

In the 1950s, shortly after they first entered Britain, I was fortunate to handle two Weimar bitches for neighbours of ours. They proved very biddable and easy to train. My abiding memory of the breed from those days was the disparity between dogs and bitches. The latter were often very sensitive, if not downright timid, while the dogs were a completely different kettle of fish. Two early dogs, Strawbridge Oliver and Sandrock Admiral (I hope I have remembered the names correctly) were absolute powerhouses and stood no nonsense. It was no surprise to find that the breed did well in other spheres than the shooting field. Several have gone on to working trial success and one dog, qualified as a police dog with the Metropolitan Police, had to be withdrawn as it became too 'bite happy'. I saw it work and fully understood the reason for such a decision.

The breed continues to be worked at Field Trials and Gun Dog Tests with great success. Again, if you are looking for one to work make sure it comes from a proven line and working parents.

Visit www.weimaranerclubofgreatbritain.org.uk

Hungarian Vizsla – smooth and wire-haired.

This is the only breed of dog which is all one colour, as the Vizsla's coat, eyes, nose, claws and pads are all the same. The coat is a copper/gold one, ranging from deep chestnut to sandy. It is an extremely old falconry breed, dating back over a thousand years. It is more slightly built than the GSP and the Weimaraner, and tends to be sensitive and clingy – to the point at which the dog is limited as a hunting dog, because it will not range out far enough to cover sufficient ground.

A brace of woodcock flushed from the one point

Experienced trainers recommend letting a Vizsla run pretty wild for the first twelve months to build its independence and initiative. Once it has gained confidence and developed its nose it will still be reasonably easy to control, when its responsive nature will help you attain excellent results. As a pointing breed Vizslas are very 'birdy'.

photos courtesy of Debby Wearing

Vizslas work in the shooting field and with falconers, and are successful in both disciplines. Opinion on the breed is divided and you either love them or dislike them according to your own temperament. The hot-headed or heavy-handed would be well-advised to look for another breed. If you are not naturally tough with a dog, the Vizsla may well flourish in your care. Some Vizslas can be aggressive to other dogs, so be quick to control this, particularly in puppyhood.

Visit www.hwva.org.uk and www.vizsla.org.uk

Italian Spinone

The largest of the HPRs, the Spinone is a rough-coated dog with a typical raking trot that enables it to cover a lot of ground with ease. As a working dog it can hardly be described as dashing, so the Spinone loses on style compared to the other breeds. It is brown or lemon, often with a predominance of white patching and a varying degree of ticking. One advantage of the paler individuals is that sheep seem unbothered by them – 'Is that a sheep grazing or a Spinone on point?'

Photo courtesy of Selina Leamy

The breed is comfortable in water, retrieves well, but is held up by heavy cover rather than being reluctant to enter it. Responsiveness is at times indifferent. We once lost one for three days, and eventually found it ambling happily along the road. It did not seem particularly relieved to be found, nor ecstatic to see us again. It did show great enthusiasm when reunited with its food bowl!

Physically the Spinone has more than its fair share of problems. The breed has a recessive neurological fault called *cerebellar ataxia* which severely affects movement and balance. Besides this it has epilepsy present and, as you might expect with so big and heavy a dog, it suffers from both hip dysplasia and *osteochondritis dissecans* (OCD). Before buying a Spinone you should really research good bloodlines, and take out veterinary insurance.

Visit www.italianspinone.co.uk

Sh Ch Gunsyn Pomata

Bracco Italiano

The Bracco is scarce in the UK with only 300 individuals in 2010. To the layman it is a smooth-coated Spinone, being similarly coloured and with the same long-striding gait. The long ears and pendulous lips hark back to Bloodhound or St Hubert Hound ancestry.

The breed club is still working hard to establish the Bracco as a working breed and any newcomer will get plenty of support from these enthusiasts. Expect health problems.

Visit www.braccoitaliano.co.uk

The Brittany

The odd one out, as it is much the smallest of the HPRs. Do not be fooled by their stature however – part of the breed description is 'maximum dog in minimum body'. You get an awful lot of dog when you get a Brit. They are not spaniels, but have been produced by early crosses between the *fougere* and setters and pointers. There is some inconsistency in size, build and colour in the breed. The Brittany can be clear-marked, lightly ticked or densely roaned, in orange and white (the dominant colour), black and white, liver and white and tricolour (black, orange and white, or liver, orange and white). Self-coloured dogs and sables occur but are frowned on. The coat is silky, but not soft, and carries light feathering.

The Brittany is very compact and well-muscled with a brisk, clipped gait. It can run endlessly and cope with any sort of terrain, and the deepest heather. Being so small it is able to 'snake' through brambles. Traditionally the breed is docked for work, but can also be bred tailless. As shooting dogs bitches may struggle to retrieve cock pheasant and hare, simply due to the size of their muzzle. Although uncommon in Britain they are the most used gun dog in both mainland Europe and the USA (where the American Brittany is a leggier, lighter-built animal to cater for the extremely vast terrain they are asked to work).

In training they exhibit a complex mix of wilfulness, exuberance and sensitivity. Often exhibiting initial reluctance to enter water, or to retrieve, they need sympathetic, patient and firm handling. As game finders they are without equal, and their drive when hunting is legendary. The working Brittany makes an exceptional family pet but, unless worked, may make your life a misery. The breed is well served by a very enthusiastic club.

Visit www.brittanyclub.co.uk

Photo by Countryshots

Chapter 2
Choosing Your Puppy

Probably, having got as far as this, you have already decided which breed is right for you. Can I beg you to rethink? You are embarking on a partnership that may last for fifteen years, and will certainly cost you many thousands of pounds in food, veterinary fees, boarding costs, not to mention destroyed furniture and possessions. You will invest long hours, and possibly change your whole lifestyle to accommodate this new tyrant who is entering your home. And your puppy will be every much a tyrant as a new baby, even if its childhood lasts for much less time.

Think also of your family. Will they be capable of controlling a large dog physically? Does your partner value his or her garden – for it will never be quite the same again? Are any of your children likely to be allergic to the type of hair your chosen dog will have, and can you trust them to close doors every time? Should you get a different car to provide dog space in the back? Are you going to depend on someone else to exercise it in your unplanned but predictable absences? How much are you really going to use the dog?

These questions are necessary as all pointing dogs are high maintenance. The pointing response is an anxious one. The dog points because it is unsure and insecure. This is reflected in your pet's behaviour in that it will be surprisingly dependent and needy. Pointing dogs crave company, and are not easy dogs to leave for long periods. They demand attention, but most of all they require exercise.

The various pointing breeds evolved to do a very specific job and, for most, this entailed working vigorously all day. It also demanded that they covered vast areas of land to find scattered game, as stocked shoots did not exist when they were developed. When I was hawking full-time over my Brittanys, I estimated that they covered between 70–80 miles per day – and did this every day for five months! Do not be surprised that your dog will still be capable of this. Unfortunately, in our increasingly crowded world there are not many people who have access to wide open spaces on a daily basis. Instead you may be forced to provide exercise on the lead, which will make you and the dog fitter, but will never tire it out. Running is what it needs and you may find it hard to budget that sort of time, but without it your dog will be boisterous, unruly and possibly destructive. Recognise that you are taking on a workaholic with bottomless stamina – and be prepared to cater for it.

Outside the hunting season I try to give my dogs at least an hour of free running daily. This adds up to the best part of one full working day each week. Allow extra for grooming and poo picking etc. and you have put in a lot of unproductive and free overtime. Have I failed to persuade you to drop this silly plan? OK, let us do it properly then.

The pointing dog as a pet

A great many pointing dogs are bought as pets by people describing themselves as an 'active family'. Usually this means that they tend to go walking, running or cycling at weekends. Their expectation is that the dog will join in with their activities. Unfortunately, the dog expects that they will want to join in with *its* activities!

What the pointing dog wants and needs to do is get rid of a lot of energy, and to satisfy its hunting urge. The urban dog owner may well find that their pointing dog can very quickly exhaust the potential of their local park. They need to find ways of getting rid of its pent-up energy in a way that makes sense to the dog. Although you may have no opportunity, or even desire, to become involved with falconry or shooting, there are elements of the working dog's skills that you can exploit to give it the sort of outlet it needs.

Retrieving is probably fundamental to this as it provides the chance to give your dog a lot of vigorous exercise in a reasonably restricted area – and in a short time. It can also be used as a reward when the dog comes back to you – so that it is coming back *for* something, rather than coming back as an end to its freedom. Do not just stop when it will retrieve a thrown ball or toy. Start throwing this into cover so that it has to consciously search for it. It will quickly learn to use its nose. Then start doing blind retrieves, where the dog has no idea, other than your command to 'Fetch', that there is anything to find. This will teach it to rely on, and seek, your directions, so it will always see you as an essential ally in its activities.

This willingness to look to you for directions will help if you decide to try **agility** or **trail-running** with your pet. Both are popular with town-based dog owners and give first-class exercise for both of you. Pointing dogs are good at both these sports, but they are not particularly exercised mentally by them. There are **scentwork** classes that are being run in some areas and online, and you should expect your pointing dog to find these ridiculously easy and enjoyable.

Obedience training is a necessary education for all dogs. Pointing dogs are unlikely to excel at any of the competitive skills, as they cannot switch off their inherited urges very easily. If possible, try to find gun-dog training classes, where the nature of the beast is better understood and catered for.

For more information and contact details visit www.thekennelclub.org.uk/activities

Sourcing your dog

Of the breeds we are dealing with very few are numerically common. This would seem to make finding one quite difficult but in fact it is easier. There are fewer breeders to choose from and therefore you can sort the wheat from the chaff more readily. Most breeders, except for commercial puppy farmers who are unlikely to supply such specialist breeds anyway, are in it for love of the dogs. They will all equally have their own particular interest, be it showing, shooting, falconry or a combination of these. You will have no problem getting them to tell you all about their dogs, and why they are superior to anyone else's. (It is called kennel blindness and is an advanced form of breed blindness, in which any fault the chosen breed may exhibit is resolutely ignored.)

Given that they are biased in their opinion it makes sense to get a second, more objective one. Another breeder cannot be relied upon to give this, as they will always be critical of other people's stock. Instead you should go to field trials and gun-dog tests, and ask among those competitors who seem to have good dogs and some idea of what they are doing. Falconers can also go through dog-using members of their local falconry clubs, or attend field meetings and falconry fairs. If you see several dogs that you like from the same kennel, then you should approach the breeder and be prepared to wait for a puppy from him or her.

Another good ploy is to identify those male dogs that seem to sire good workers. They were used as mates by knowledgeable people who appreciated their good qualities and bloodlines. Exploit this wisdom and ask the stud dog owner if there are any good working bitches about to produce pups to him. They, of course, want their dog's offspring to be worked, and they will guide you towards a suitable litter.

Check for hereditary health problems

Before committing to buying a puppy, ask the breeder about any health problems that are within the breed, and whether the parents of your puppy have any record of these problems among their close relatives. Breeders should not object to these questions, as they are the same ones they will themselves have asked before planning the mating. If you meet with a lack of openness you should think twice. For a second opinion contact the breed club's Health Representative (all clubs are required to nominate such an officer who submits an annual report to the Kennel Club). Responsible breeders will be guided by the Kennel Club's Assured Breeder Scheme or by the breed club's Guide to Ethics.

Hereditary health problems are normally, but not necessarily, associated with recessive genes, and any dog may have them lurking somewhere in the background of its ancestry. Because of this they can occur at any time but, by choosing from stock that is in itself vigorous and healthy, you are minimising the risk. Remember most dogs are healthy and fit, and this is as true for pedigree dogs as for any street mongrel. Indeed, the term pedigree only means that the dog is of known breeding for five generations and, as we have seen with our more recently manufactured breeds, they often come from a judicious mix. A breed is the result of a conscious, planned selection of desirable physical, temperamental and behavioural characteristics. The mongrel is only selectively bred to roam, as this is by definition the one trait all its ancestors share.

Hip dysplasia (HD) is the most prevalent condition that is routinely checked, and it is assessed by X-ray examination. The British Veterinary Association, in tandem with the Kennel Club, issue certificates for screened dogs. Breed clubs will tell you the average score for their breed. However, most dogs have a degree of dysplasia, and to some extent you should consider its presence to be normal. At an acceptable level the dog will function perfectly well for a full working life but, as a general rule, the heavier the breed the more serious the implications.

Brittanys rarely suffer any functional problems because they are small, well-muscled, very compact and run on their forehand. This term means that they are front heavy when running, and the hind end supplies limited propulsion and has little to do but keep itself up. In contrast, breeds like Spinones and Braccos are big, not at all compact in movement, and place more weight on the hind end.

Where an excessive degree of HD is present there will be consequences for the dog in later life. Depending on the severity of the condition the dog will experience discomfort and arthritis, and may become inactive prematurely.

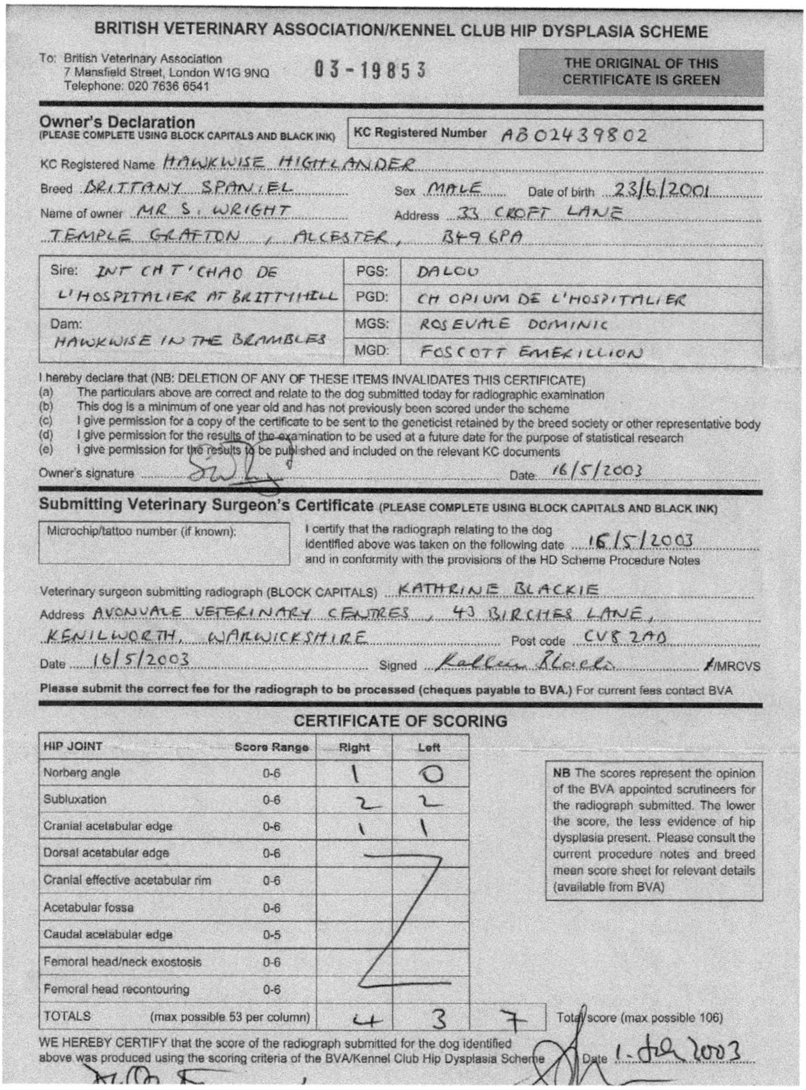

BRITISH VETERINARY ASSOCIATION/KENNEL CLUB HIP DYSPLASIA SCHEME

To: British Veterinary Association
7 Mansfield Street, London W1G 9NQ
Telephone: 020 7636 6541

03-19853

THE ORIGINAL OF THIS CERTIFICATE IS GREEN

Owner's Declaration
(PLEASE COMPLETE USING BLOCK CAPITALS AND BLACK INK)

KC Registered Number AB02439802

KC Registered Name HAWKWISE HIGHLANDER
Breed BRITTANY SPANIEL Sex MALE Date of birth 23/6/2001
Name of owner MR S. WRIGHT Address 33 CROFT LANE
TEMPLE GRAFTON, ALCESTER, B49 6PA

Sire: INT CH T'CHAO DE L'HOSPITALIER AT BRITTYHILL		PGS:	DALOU
		PGD:	CH OPIUM DE L'HOSPITALIER
Dam: HAWKWISE IN THE BRAMBLES		MGS:	ROSEVALE DOMINIC
		MGD:	FOSCOTT EMERILLION

I hereby declare that (NB: DELETION OF ANY OF THESE ITEMS INVALIDATES THIS CERTIFICATE)
(a) The particulars above are correct and relate to the dog submitted today for radiographic examination
(b) This dog is a minimum of one year old and has not previously been scored under the scheme
(c) I give permission for a copy of the certificate to be sent to the geneticist retained by the breed society or other representative body
(d) I give permission for the results of the examination to be used at a future date for the purpose of statistical research
(e) I give permission for the results to be published and included on the relevant KC documents

Owner's signature [signature] Date 16/5/2003

Submitting Veterinary Surgeon's Certificate (PLEASE COMPLETE USING BLOCK CAPITALS AND BLACK INK)

Microchip/tattoo number (if known):

I certify that the radiograph relating to the dog identified above was taken on the following date 16/5/2003 and in conformity with the provisions of the HD Scheme Procedure Notes

Veterinary surgeon submitting radiograph (BLOCK CAPITALS) KATHRINE BLACKIE
Address AVONVALE VETERINARY CENTRES, 43 BIRCHES LANE,
KENILWORTH, WARWICKSHIRE Post code CV8 2AD
Date 16/5/2003 Signed [signature] MRCVS

Please submit the correct fee for the radiograph to be processed (cheques payable to BVA.) For current fees contact BVA

CERTIFICATE OF SCORING

HIP JOINT	Score Range	Right	Left
Norberg angle	0-6	1	0
Subluxation	0-6	2	2
Cranial acetabular edge	0-6	1	1
Dorsal acetabular edge	0-6		
Cranial effective acetabular rim	0-6		
Acetabular fossa	0-6		
Caudal acetabular edge	0-5		
Femoral head/neck exostosis	0-6		
Femoral head recontouring	0-6		
TOTALS (max possible 53 per column)		4	3

7 Total score (max possible 106)

NB The scores represent the opinion of the BVA appointed scrutineers for the radiograph submitted. The lower the score, the less evidence of hip dysplasia present. Please consult the current procedure notes and breed mean score sheet for relevant details (available from BVA)

WE HEREBY CERTIFY that the score of the radiograph submitted for the dog identified above was produced using the scoring criteria of the BVA/Kennel Club Hip Dysplasia Scheme Date 1-Jul 2003

British Veterinary Association and Kennel Club Hip Dysplasia certificate

Osteochondritis Dissecans (OCD) is a bone condition affecting the shoulder and foreleg, and is primarily seen as front end lameness in growing youngsters of the heavier breeds. Like HD it is multifactorial and is often associated with precocious growth of heavy bone. It frequently rights itself as the dog matures and the soft bone hardens. The condition is aggravated by breeders feeding their puppies excessive, or over-nutritious, diets to promote early substance. Exercise at this age must be restricted to minimise damage to the elbow joint in particular.

Progressive Retinal Atrophy (PRA) or night blindness was particularly common in Irish Setters. Fortunately, the inheritance of PRA has been understood for some time, and the breed club has achieved wonders in identifying the carriers and in eliminating it. One can never say it may not occur again but it is now a rare condition in the breed. When the condition occurs the blood vessels in the retina wither slowly away so that the more peripheral cells die off. These are the most light-sensitive and the dog will function less well in duller conditions. I have seen a dog hesitate as it came to shadows of trees that crossed the pavement it was walking along. In advanced cases the dog becomes functionally blind at night, and eventually will lose all its vision.

Epilepsy seems to occur in a few pointing breeds. Possibly this is associated with their hyperactive natures and inbuilt anxiety. At its least noticeable level it is little more than an inconvenience to the dog, which may exhibit mild shivering or teeth chattering. Severe cases will involve convulsive behaviour and loss of consciousness. In the first instance the condition can be readily controlled by administration of Epiphen, or some similar drug, and the dog will lead a perfectly normal life. Extreme cases may require euthanasia.

I had a Brittany bitch which exhibited teeth chattering. A sharp word was enough to stop this but we identified it as epilepsy (without resorting to expensive brain scans), by simply administering one Epiphen tablet a day. This stopped any recurrence. Because I had originally bought her for breeding, she was no use to me, and I gave her to a friend who worked her successfully for the rest of her life. He actually discontinued the tablets and simply controlled her attacks verbally. (NB Before transferring her I put a restriction on her registration to prevent her ever being bred from.)

Around the same time, I was lent a Brittany brood bitch to work for a season. Within an hour of getting her home she threw a severe fit in our lounge. She bit her tongue quite badly, and was also doubly incontinent. I persevered for a couple of months, but she fitted violently about twice a week. The final straw was when she suffered a fit whilst working a rocky hillside in Scotland. Losing all balance, she cannoned off a succession of boulders and eventually ended upside down in an ice-cold burn. Fortunately, this short, sharp shock brought her round. What worried me most was that her owners were unaware of her epilepsy, as she had been kennelled and used just for breeding. They evidently spent little time with their dogs. (There were quite a few epileptic Brittanys imported in the 1980s, but the breed club quickly made this public and published the pedigrees. In consequence the problem is now almost unheard of in the breed.)

Epilepsy can normally be controlled by administering inexpensive drugs, so it is not a problem in a working dog. However, such an individual should never be bred from, and both the breeder and the Breed Health Representative should be notified.

Cerebellar Ataxia (CA) is a neurological condition which affects the balance and locomotive skills of young Spinones. This is caused by a recessive gene which has been identified as being solely inherited from one imported dog. This may be incorrect if any of his direct offspring were affected, as the bitches to which he was mated must also have carried the gene for it to have become evident. Affected stock will not live long enough to be bred from (the condition is visible at an early age) but carriers may not be detected for generations. As the original dog inherited CA himself, so his ancestors will also have spread it through the breed. It is hoped advances in DNA analysis will help identify the faulty gene and lead to its eradication.

Hereditary Cataracts affect Large Münsterländers. Like a lot of minority breeds one fault introduced by an imported stud dog can be disastrous. Being of fresh bloodlines a lot of bitches will be served by it, and its influence will spread rapidly throughout the breed. Subsequent matings of related stock will exacerbate the problem, which the original dog may not have exhibited. This happened in the 1950s in Boston Terriers when an imported dog – Acecourt Wee Michael – was brought in to add variety to the gene pool. He was used extensively but was found, too late, to be a carrier of hereditary cataracts. It took a long time to eradicate this condition from the breed. The Large Münsterländer Club are working hard to identify the faulty gene.

Dog or bitch?

The qualities we look for in a pointing dog depend on the individual animal, irrespective of its gender. Both dogs and bitches can have superb noses, bottomless stamina, responsiveness and a hunting focus. Equally there will be as many natural 'retrievaholics', water lovers and 'cover crushers' amongst either sex. These are virtues that have been bred for and should be present if the breeder has got it right.

There are plenty of vices also, which either sex may exhibit. Gun shyness is an obvious disaster for the shooting enthusiast, and nobody wants a hard-mouthed, chasing killer. Failure to point or retrieve is just as unforgivable, as is a dog which is too sensitive to enter water or heavy cover. However, falconers do not mind dogs that will not retrieve or swim, and the sound-shy dog gives them much easier control.

Is there any benefit or advantage in one sex over the other? I suspect much of the debate is due to prejudice. I have always thought of myself as a bitch man, having only owned five entire males compared with twenty-four bitches to date. Yet, when I think of those five dogs, I have to say they were all a pleasure to own and absolute gentlemen.

For me part of the attraction of bitches is that I like breeding. If a bitch fails to reach the highest standard you can still mate her to the best dog in the world, and have the pick of the resulting offspring. By contrast, if your dog is average, nobody will bring the best bitch in the world to him. Therefore, it is easier to develop a quality kennel from bitches. The downside is that bitches come in season every six months, which is inconvenient if this falls in the middle of the shooting season. While she may still be prepared to work in that condition her presence might prove an unfair distraction for any males in the field! Your vet can inject your bitch to delay the onset of her season, but this will mess about with her hormones and may adversely affect her fertility if you intend to breed from her in the near future. Unless you have a pressing reason to do this you are advised to avoid any unnecessary use of chemicals.

On the subject of sex, it must be mentioned that male dogs usually have a healthy interest in this, and may be prone to wandering in search of an accommodating female. Nobody wants their dog to cause a road accident. It is the responsibility of every dog owner to make sure their pet is within a secure garden or run. Check the fencing before buying your pup as, if it learns to get over, under or through fences at an early age, it will always try to find a weak spot.

Both sexes can be neutered if you have no interest in breeding. In males this should be left until the dog is regularly cocking its leg, and bitches should not be spayed until after they have had one season at least. Castration is likely to reduce the incidence of aggression in a male and make it much more biddable. Neutering a bitch is a convenience, and also helps protect her from a septic womb in later life.

Ironically, at a time when many vets are vocal in their condemnation of docking as a form of mutilation, most of them are more than happy to neuter dogs. Given a choice between no tail and no 'bits' I am pretty sure which way most of us would choose. More worryingly, in minority breeds, this reduces significantly the available gene pool. Having received several requests from people wanting to use a son of my old stud dog, I contacted all the owners of my dog puppies. Of these almost two thirds had been castrated!

Temperamentally there is more likelihood of an entire (uncastrated) dog requiring vigorous control than a bitch, and also of exhibiting aggression towards people and other dogs. They are larger and stronger, so you need to consider the control capabilities of the younger or older residents in your house, before foisting a powerful delinquent on them. Avoiding problems by anticipating them is much better than having to deal with them once they occur, and we will hopefully cover this in subsequent chapters.

A little-mentioned aspect of working bitches is that their concentration and drive will be affected by their menstrual cycle. They will become increasingly energetic and excitable as they approach their season, giving rise to a need for greater control. Once the season has passed the bitch may then become much quieter and sedentary, and will appear to lack a little enthusiasm and drive.

This phase can last for about fifteen weeks (about the length of time she would be carrying and subsequently nursing her litter). Only after this time will she return to a normal level of activity before peaking again a couple of months later. The degree to which this variation occurs will be different from bitch to bitch, but the ones where it is most clearly evident will suffer false pregnancies, and may need spaying in later life to avoid septic wombs.

Buying two or more puppies

Very occasionally people buy more than one puppy from a litter on the basis that they will be 1. company for each other, 2. you will have the chance to choose the best when they are further along in their development.

Please do not be tempted to fall into this trap. They will have their own little community, so will not be as centred on interacting with you. One pup at a time is enough to concentrate on. It is also very difficult to keep a dog and bitch together – if nothing else it can be a logistical nightmare keeping them separate when the bitch's inevitable seasons come round.

Booking a puppy

If you have planned ahead properly you will have sourced your puppy before it is born. It is quite reasonable for the breeder to request a deposit, as otherwise they can get an awful lot of time-wasters filling up their order book. It is also an advantage to you as it ensures you will get your pup.

My own procedure is to accumulate names and addresses of prospective purchasers in the months between litters. At the time of mating I then inform all interested parties, so that they can confirm their continuing interest. Once the litter is whelped, and safely docked, I know how many pups I have to sell.

Again, the potential buyers are contacted, in chronological order from the date they first contacted me and expressed a firm interest. They know how many dogs and bitches the dam has had, and what colours and markings are available. By sending a deposit they secure their particular place in the queue, e.g. second pick of the bitches or fourth pick of the dogs. Everybody must wait until those before them have chosen.

At what age should you buy your puppy?

Guide dog organisations throughout the world have proved that placing their puppies in homes at six weeks produces the best temperamental development. In the UK alone that equates to around a thousand pups per year over more than half a century – they probably know what they are doing! Despite this it is now illegal under the Animal Welfare Act 2006 to sell pups until they are eight weeks old.

This is potentially damaging to the puppies' social development and I try to get them to their new owners with as little delay as possible. It takes some arranging but I expect to have a succession of buyers arriving every couple of hours. My last litter of nine were collected over the space of just two days. Unless pups are given individual attention, they will soon become pack animals, and will never quite be the family dog that you want.

What you should get with your puppy

Any breeder worth their salt will furnish you with the following:

- A five-generation pedigree including the Hip Dysplasia Score of both parents, and any other health screening results. (*I also provide photos of ancestors where I can, as a pedigree is otherwise just a list of names*).
- A Kennel Club Registration document that he/she has signed to enable you to transfer the puppy to your name. (*Some breeders impose breeding restrictions to prevent you breeding off your dog without their consent. They believe they have the knowledge, and should have the right, to dictate what stock is bred from in the breed. This is a disgrace. The only person who knows your dog's virtues and weaknesses is you. It is your decision to breed from your dog. By all means take advice before committing to it but remember it is your choice.*)
- A Certificate of Legal Docking, signed by a qualified veterinary surgeon if applicable.
- A microchipping certificate and registration document, signed at the same veterinary practice if the puppy is docked.
- A diet sheet for your puppy (with sample food for the first couple of days), some advice on toilet training, grooming, exercise, socialisation and on when it has been wormed.
- Temporary Health Insurance cover supplied by the Kennel Club or microchipping agency.
- His or her contact details for advice or in the event of the puppy needing re-homing.

What to look for when selecting a puppy

Health must be the first consideration. The youngsters should be bright-eyed and alert with supple skin and a well-to-do air about them. Their nose and eyes should be free from any discharge. When picked up, the youngsters should feel surprisingly solid.

A well-nourished litter will offer plenty of evidence that their intestines are functioning well. Puppy faeces are often soft but should not be sloppy, liquid, bloody or frothy. Any sign of them dragging or 'scooting' their backside along the ground can hint at worm infestation. This is not the end of the world but it does call into question the expertise of the breeder.

It would be nice to see the pups being fed as they should attack their food with relish. However, a helpful breeder will have delayed feed time, so that your puppy will not have been fed – and is therefore less likely to suffer travel sickness on the way home.

Let us suppose that you have the pick of the litter and have no preference for either sex or colour. You may be faced with eight puppies, four of each sex, and have to make your mind up. What will help you choose at this age?

Eventual size is usually reliably indicated in a well-reared litter. Dog pups will bigger on average than their sisters, but big females may be larger than some of their smaller brothers. Some will have bigger feet and more pronounced 'knuckles' on their front legs, indicating that they will grow on and have heavier bone. Any lacking these indicators will probably grow up to be light-boned and may be weedier (but possibly more agile).

These youngsters are chunky and well-fleshed, with bright eyes and healthy coats

If you are looking for a dog to do a lot of work the presence of very heavy bone may be a disadvantage. A working dog must have the correct power/weight ratio and, provided it has good straight legs, a lighter-boned dog may have an advantage – displaying greater stamina and agility. Talk to the breeder about this if you are unsure. In a litter that looks underfed or 'wormy' (i.e. not well-fleshed but pot-bellied) the pups will not have had a good start. They may have a greater potential for growth so it is harder to judge their eventual size at this age. Hopefully you will not be looking at this sort of puppy anyway.

Temperament can often be accurately gauged. Firstly, you have already got a fair idea about the parents from your earlier research. Is the bitch still happy with the litter or does she avoid them? Was she protective or nervous when you visited? A few warning barks is natural, expected and totally acceptable, but you want her to settle down quickly and greet you pleasantly.

The pups will respond to your arrival. In a litter that is handled a lot, as they should be, you can expect them to overwhelm you in a tidal wave of enthusiasm. Do not wear good clothes to inspect a litter! Better still be prepared to sit on the floor with them.

A good litter should clamber all over your head, pulling at your bootlaces and chewing your hands. Some individuals may be less forceful and demonstrate a more sensitive nature. Provided they do overcome this there is no reason why they should not be perfectly good workers – and easy to control. In contrast the pushy pups may be ones that you will always have to be on top of.

The ideal response – confident and enquiring

I always look for the pup that makes eye contact with me. This awareness of you is a prerequisite of attentiveness and will make training much easier. It can be taught, of course, but some pups have it in a larger share than others. Make a little noise, perhaps clicking your tongue, to see if it cocks its head to one side, or comes up to find out what you did. Talk to it and see if it listens. Remember it will have a limited concentration span so don't be too ambitious.

If your dog is intended for the shooting field it makes sense to clap your hands loudly to gauge its reaction to sudden noises. Talk to the breeder too about how he or she perceives the individual natures of the pups.

In my last litter I had two very nice pups but one was hyperactive, and so confident that I swear she had been here before. She needed a knowledgeable home where she would be stimulated and given a lot to occupy her. Pay her any attention and you were immediately engaged in a mad ecstasy of affection and playfulness. Her brother, in contrast, was a sweet-natured, placid pup who wanted nothing better than to lie against your feet. Make a fuss of him and he would just roll over and wait for his tummy to be tickled. Both would be suitable for work, but this bitch would relish a real workload whereas her brother would have suited a weekend shooter. Remember that your puppy is going to be half of a team. You are the other part. A little self-knowledge will help you choose a pup that will suit your temperament. Be honest with yourself and recognise if you tend to be heavy-handed, or if you are likely to spoil a puppy and make a pet of it.

If you fall into the first category you might be well advised to pick a 'rough toughie' who can bounce back after a telling-off. However, those of us who favour a softer approach will find that a more sensitive soul might blossom in our care. Of course, you may not have much of a choice so be prepared from the outset to modify your handling style to suit the pup.

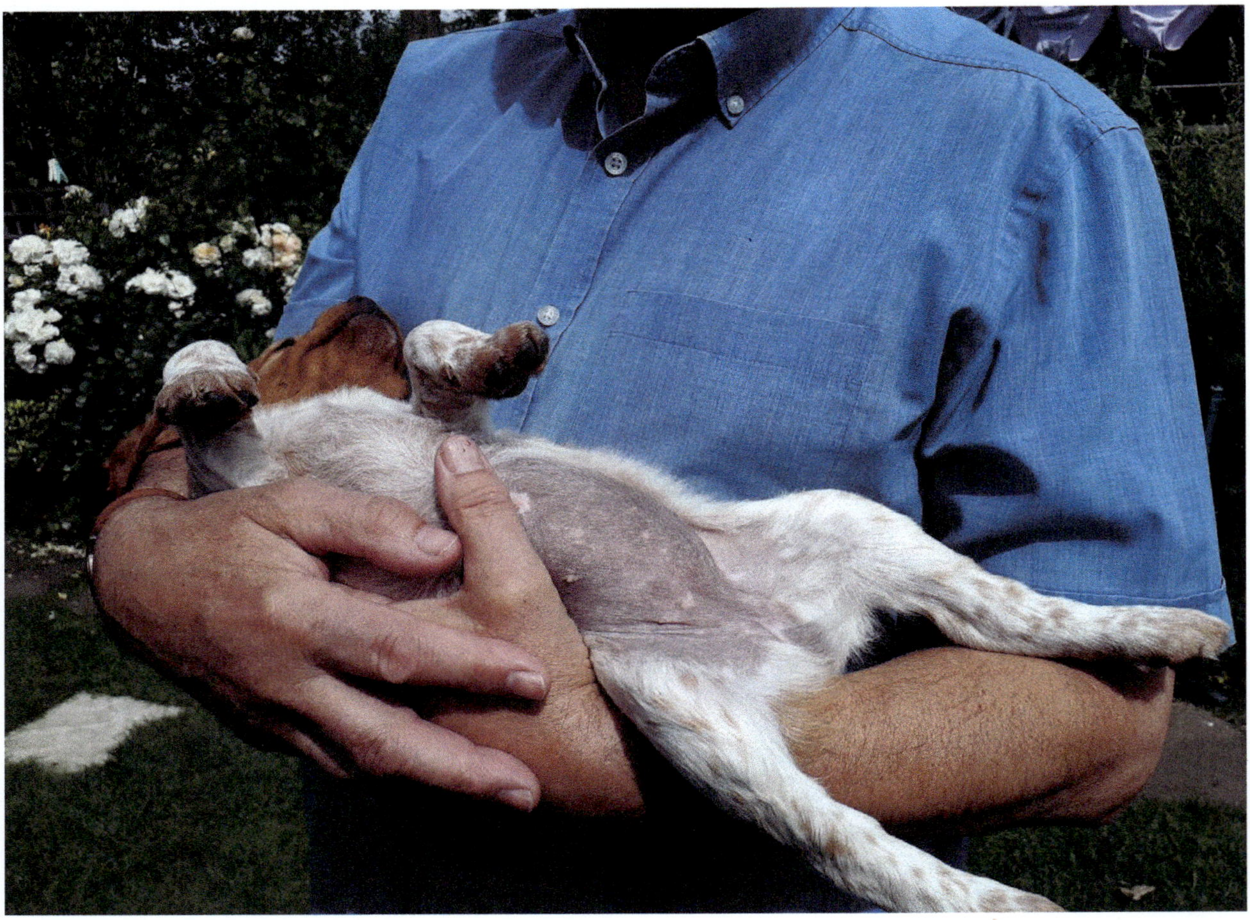

A good indicator of physical and temperamental confidence is to roll a puppy on to its back. One that is relaxed in this position is definitely not nervous!

Travelling your puppy home

Being in a moving environment can be a scary experience for a puppy. Until now it has been the one who moved while everything around it stayed still. Its eyes are able to focus on passing scenery, but it is being bombarded by a myriad new sights at a speed it cannot understand. What it needs is comforting, so it is a good idea to take someone with you to act either as nursemaid or chauffeur. The puppy will be happier if it is cuddled and can chew someone's fingers to distract it from the mad rush that is passing the car window. Make sure you have plenty of old towels and kitchen roll to mop up any vomit or saliva.

If you are alone you should have a pet carrier with 'Vetbed', towels or an old jumper for the pup to lie on. Ideally belt this securely on the passenger seat so that it can lick your fingers through the mesh. It will really need human contact and affection at this time.

First signs of trouble are the puppy licking its lips and possibly yawning. This is followed by an increasing amount of bubbles and froth around its mouth. Pretty quickly it will throw up. Once this is done it will probably settle down and may well go to sleep. Do not ignore the problem as a travel-sick dog will be an inconvenient companion at the very least.

You can ask your vet for travel sickness pills or mild tranquilisers when travel sickness persists. Opinions are split as to whether it is better to take long trips so that the pup gets resigned to it, or to stick to short journeys to the park so that it does not have time to get sick. This gives the pup the idea that driving anywhere is exciting and rewarding. I have tried both systems successfully so you must be guided by the results you get.

Settling in

I assume that you are going to keep your pup as a house pet, although many professional gun-dog trainers and gamekeepers advocate keeping working dogs in kennels. Their reasoning is that this stops you being soft on the dog, thereby letting it develop bad habits and a lack of respect for you. Equally their own dogs, which only get attention when being trained or worked, put all their energies into their work. They regard the system of kennel management as integral to this. We will come back to this when we begin to train our puppy but, for now, I assume you are going to have a family pet, as a single dog will be unhappy kept by itself in a kennel.

Give it lots of affection so that you become its sanctuary. Particularly keep taking it out into the garden so that it has the chance to relieve itself. There will be mistakes for sure but, if you are zealous in avoiding them, it will help establish good patterns of toilet behaviour from the start.

To help the puppy feel at home you should feed it as soon as it has explored its immediate surroundings and has begun showing signs of increasing confidence. The breeder should have given you a sample of its normal food to use for the first day or so. Having come from the competitive situation of a litter it will eat readily to start with. You may find that it gets less greedy in time, in which case you can begin to reduce the amount or frequency of its meals. For the first few months it should retain its puppy chubbiness – until its legs start to sprout. It is like a teenager going gangly after having been a plump infant.

When you arrive home with your puppy it will be frightened. It will miss the support of its litter mates and everything will be new. Expect it to be subdued and to examine each new thing with great caution.

Ask its breeder for a piece of used blanket or Vetbed to put in its crate – this will hold the scent of its brothers and sisters. This will be comfort to it at night. Also give it a soft toy or something to play with and cuddle up to.

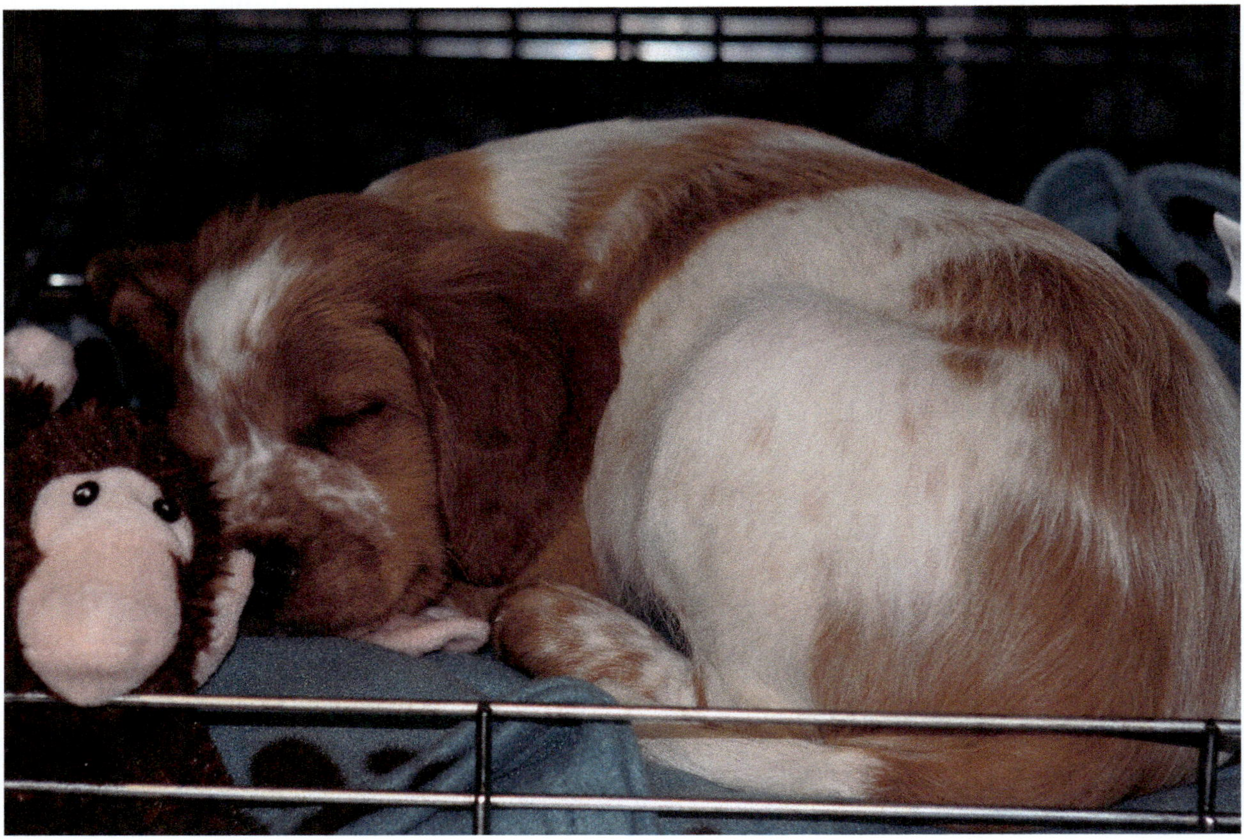

Photo by Caroline Perks

Introducing the puppy to other dogs

If you already have an older dog you should take care to supervise their meeting, having first put the established dog on a lead. The puppy will expect any adult dog to be a mother to it, and will approach with complete lack of fear. It may be over-confident, jumping up to suckle or play. Not all old dogs like pups, and they may resent an intruder in their settled world. Sibling rivalry and jealousy are not restricted to children. Telling the older dog off is not always the best solution, as it may harbour a grudge. It is important that its status as a senior dog remains intact, and any discipline by you will undermine this. You are much better off trying to restrain the pup while reassuring the old dog that it is still special.

With time, and by making an effort to give the older dog plenty of attention, it will accept the young upstart. Dogs are by nature social animals – and it is within their scope to accept additions to their pack. They may discipline it much harder than you would, but it is rare that a puppy gets injured in this situation. Indeed, you can watch and learn. Their control can be more severe than you would use, but the puppy accepts it and an unquestioning hierarchy is established.

You should never give the puppy priority treatment, or be seen to favour it. This is especially true with food. Put the old dog's food bowl down first, then the pup's. Give the old dog a titbit when the pup is having extra meals. Only when the two play together, and the older animal is displaying good tolerance, can you feel confident that a good relationship has been developed. Until then never leave them unattended together. It may be necessary to shut the puppy in its crate to keep it separate from the older dog.

Introducing to pet cats

By and large your cat will simply refuse to acknowledge the presence of the new puppy. Unfortunately, this will not deter the pup from trying to make friends with this strange 'dog'. If your cat just votes with its feet and keeps out of the way, you can relax but, if it stands its ground, be very careful. The pup has little sense of danger and will not read signals such as an arched back and hissing. Naively it may continue to investigate and be rewarded with a swift swipe from a clawed foot. Nothing is guaranteed to teach it a quicker lesson but it can be an expensive one for you. A few drops of blood on the pup's nose are fair enough, but cats can aim pretty well. It is not unknown for Puss to scratch the pup's eyes. You really do not want a vet's bill on the pup's first day in your home, so monitor their relationship with great care.

The first night

Do not expect your puppy to be quiet when left. Very possibly it will whine and wail for company. Ignore it for as long as you can – it may give up first. If it persists you can tiptoe downstairs without it hearing you and interrupt it in mid-whine. Sharp words and a quick shake by the scruff of its neck should suffice to shut it up for a while. A determined pup might need this sort of treatment two or three times until it learns to keep quiet at night. The one thing you must not do is give in to it and take it up to your bedroom.

Crating will help the puppy's social behaviour

I am a firm believer in the value of introducing the puppy to an indoor kennel or crate of some description. It is always useful to be able to contain it, and familiarisation with its kennel should start from the outset. Introduced at an early stage, the crate becomes its sanctuary and escape.

If you have young children who may pester the puppy into a state of exhaustion, they can learn that it ***must*** be left alone when it puts itself to bed. Puppies have rights too!

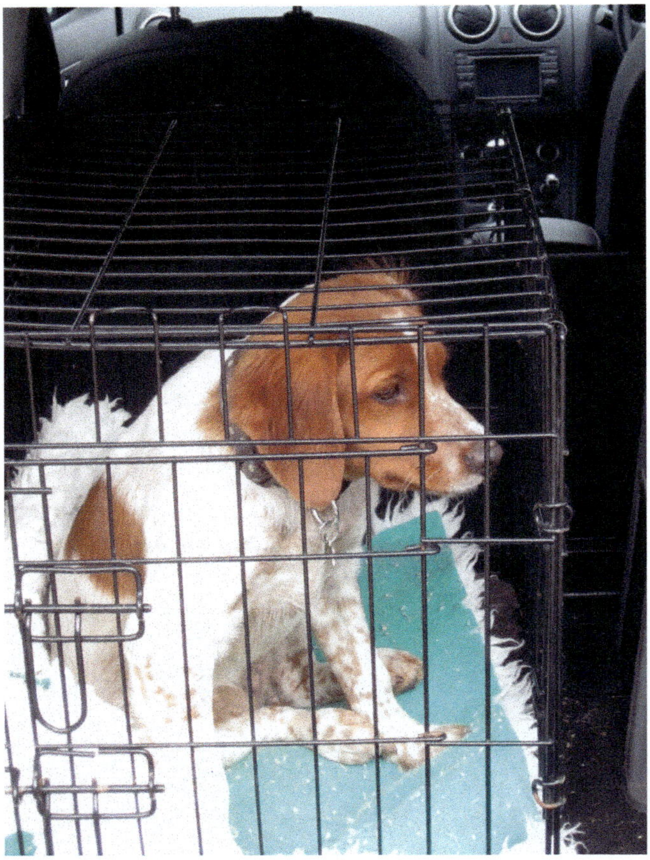

I was once asked to take in a young Boxer dog, after it had badly bitten one of the owner's children, while they decided what to do with it. When it arrived it seemed perfectly biddable, and was put in a warm kennel with some food, where it promptly curled up and went to sleep. It did not wake up for another 36 hours! The poor beast had been subjected to endless pestering until it had become totally desperate

A further benefit is that when you travel anywhere, your puppy will travel confidently in its crate. Once you have arrived at your destination, it can safely be taken into a hotel bedroom or guest house, where the crate provides a familiar environment.

Toilet training

Particularly with a new puppy, the crate also aids toilet training. No dog likes to foul its own bed and, provided it has emptied itself last thing at night, your puppy should stay clean until the following morning. Take it outside immediately and it will perform to your command. Give it plenty of praise, repeating the command as you do (I use 'Be quick!'), and it will soon get the message.

Always try to anticipate the workings of its bladder and bowels so that you can train it by positive reward. Telling it off for making mistakes will only make it very aware that its natural functions are naughty, and it will try not to let you see it doing its business. Instead it will go behind the settee, in the front room, and leave little piles there for you to find later. In its mind it is trying to be clean by choosing the quietest, least noticeable place to do it.

The dog's toilet habits are the first thing it learns, and they are the hardest thing to change as a result. If it associates relieving itself with punishment, it will be very reluctant to do so in your presence. This can be a nightmare if you are away from home and have to get it to go while it is on the lead.

The positive approach is just to get the pup outside the minute it wakes up, as soon as it comes out of its crate or immediately after it has eaten. Use a command ('Be quick!') over and over again until it obliges. It will need to empty itself anyway so you can immediately praise it, including the command in your praise, e.g. 'Well done, good dog,' 'Be quick,' 'Clever dog." You will be surprised how quickly it will fall into the routine.

Among the hang-ups suffered by dogs that have been told off for fouling in the wrong place, is a reluctance to go anywhere but in the approved place or on the normal surface. If it only goes on grass at home, it may refuse resolutely to perform on concrete, tarmac, etc. Therefore, make sure you encourage it to use different places and surfaces.

Happy to perform on the lead, and unbothered by being on a different surface. Our pups will use newspaper, concrete and grass before being sold.

Toilet training the owner

It should not need to be said but you must plan how to dispose of faeces. You may be lucky enough to have an outside toilet, or even a drain cover into your sewer that can be lifted easily. Then it is simple just to empty the shovel into it, followed by some diluted disinfectant (drains dealing just with surface water can be used, as animal faeces constitute biodegradable material, but you should flush with plain water). Lacking this sort of opportunity, you may have to install a dog toilet in the garden, which will break down faeces. These never seem to be wholly successful despite the maker's claims. Alternatively store the collected faeces in plastic bags (double-bagging is thoughtful for the bin men) and dispose of them in the general-waste wheelie bin (according to the advice of my local council anyway).

All dog owners should now be aware of the need to pick up faeces if your dog fouls any public place. Carry poo bags with you whenever you take your dog out. There may be dog litter bins available. If not, you must take your trophy home with you for disposal. Some landowners are content for you to just flick the offending poo off the footpath with a stick, where it can decompose naturally. This may be preferable to the unsightly presence of plastic bags, full of unwanted trophies, that are left hanging on bushes and fences. When in doubt pick it up!

Involuntary incontinence

A lot of young pups will urinate slightly through excitement or submissiveness – particularly when they first greet you. Being cross will only aggravate the problem so ignore it. The puppy will grow out of it.

Veterinary check-up

Within a couple of days of buying your puppy, book it in to your veterinary surgery for a health check, and for the first inoculation. The vet will listen to its chest for any irregularity in the heartbeat, take its temperature to ensure it is not brewing an infection of some sort, and palpate the abdomen for any signs of discomfort. He or she will also carry out the same clinical check that you will have been doing unconsciously: noticing bright eyes, clean breath, healthy coat and plenty of energy. All being well the vet will then give the first inoculation, which offers initial protection to the puppy against distemper, leptospirosis, hard pad, hepatitis and parvovirus. These highly contagious and potentially lethal diseases are seldom met with nowadays, because the majority of responsible pet owners do have their dogs inoculated. However, do not be complacent as they are still present, particularly in urban open spaces, and can still be contracted.

This first injection extends the pup's immunity that it inherited from its mother. In a minute doses it introduces the diseases in either a live or modified form, so that the pup builds up its own resistance to disease. After a gap of two weeks (occasionally as many as four in some vaccines) a second injection is given that further challenges the puppy's immunity and creates greater resistance. Only after a gap of a further two weeks following this injection, i.e. when the puppy is twelve weeks old, can it be considered safe to take it into public places. If you live in an area where diseases are rife the vet may even advise you to extend this period. An annual booster injection is then recommended to ensure continued resistance is maintained.

Re-vaccination

Many owners are reluctant to boost their pets' immunity at annual intervals, through concerns with the overuse of drugs of any kind. Instead your vet can take a blood sample which will reveal the 'titre' count – the level of natural antibodies that the dog already possesses. Where this is sufficiently high no booster will be needed.

Worming

Most puppies harbour canine roundworms while they are with their mother. This is because the change in her hormonal levels while she is pregnant activates any dormant roundworm eggs that may be within her body. No matter how often she is wormed (at least every three weeks throughout her pregnancy) there remains a strong possibility that her litter will also have worms. Breeders should therefore worm the puppies at regular intervals from the age of three weeks.

One of the reasons that it is necessary to worm dogs is that roundworms eggs can easily be passed on to children, who can develop a condition called *toxocariasis*. The worm larva develops inside the body and migrates to areas where there is a plentiful blood supply. This can include the eyes, where *Toxocara* has been known to cause blindness.

The anti-dog lobby is quick to jump on this particular bandwagon, but without much cause. For the eggs to be at a stage in their cycle where they can hatch inside the host's body, they would need to have experienced prolonged temperature and humidity levels that are virtually unknown in temperate Britain. The problem however is a very real one in tropical zones and is particularly associated with poor hygiene. Incidentally the cat and fox roundworms, which are just as common, can cause exactly the same affliction – and these host animals are rarely wormed.

In the 1950s a survey was carried out among dog breeders, who are obviously the people most often in contact with canine roundworms. Blood tests revealed no difference in their exposure to *Toxocara canis* (the dog roundworm) than the rest of the population.
Your vet will be happy to supply wormers to you on your first check-up visit. As the dog grows you can reduce worming to every three months. Because working dogs come into contact with farm and wild animals, they are likely to pick up worms also. These may not just be roundworms, but tapeworms and lungworms as well. Worming is only one part of the problem as the flea is the tapeworm's intermediate host. You will need to regularly de-flea your pet as well. Because of the rise in the incidence of Lyme's disease, which is transmitted by tics, it is important to use a flea treatment that covers tics as well.

Aggression

It seems strange to talk about aggression when we have only just got the puppy home. Normally any youngster is going to be playful, friendly and unlikely to demonstrate anger. However, within the litter, they will already be competent play-fighters and some do get to be quite 'spiky' with each other. Naturally, when they no longer have litter mates to play with, they will happily nominate you and your family to fill this role. Play-fighting is an essential way of developing their locomotive skills and physical co-ordination, and should not be completely discouraged. Very quickly this interaction builds a close bond with you as your pup identifies you as a good playmate.

However, some pups get a little overexcited and carried away with things. It is easy to laugh at the infant tantrums, and pick up your opponent to cuddle the aggression out of it. The family might be called to watch this ridiculous display of temper, but it would be wrong to ignore it totally. Ideally you can restrain the pup, very often to its evident fury, until the outburst subsides and you can return to more gentle play. The times to worry about are when the level of aggression recurs rapidly, and in different situations. This can be over food, or when you are holding the puppy for de-fleaing or some other indignity.

Even at this stage the puppy is socially aware. If it gets away with growling at, or even biting you in earnest, you can be sowing the seeds of serious challenges in the future. It is much better, if you judge that action needs to be taken, to be over-aggressive yourself. While you do not want to hurt the pup in any way, you do want to frighten it and give it a sense of respect for your authority.

Take a leaf out of the canine manual on puppy training. Older dogs will be very tolerant of bumptious puppy behaviour, and often allow youngsters to chew their ears, hang off their tails and generally make a nuisance of themselves. This torment is acceptable to them while the pup remains firmly in the kindergarten category. However, when they judge the pup is getting too bossy and confident, they will turn and really appear to beat it up. Their apparently savage response will leave the pup quivering in a corner for as much as twenty minutes or so. Examination will reveal no actual wounds on its body – but its self-image has been severely mauled. Having been tolerated as a puppy it has moved into the adult world and has learned to respect authority.

The probability is that you will not have to tackle this issue. It only becomes necessary where baby temper tantrums get more serious, perhaps over food, a toy or a bone. If you need to take action, do what an older dog would do. Grab it by the throat and pin it to the ground, snarling and shaking it. The puppy will tell you if you have been effective. If it is badly upset, and too frightened to come to you for reassurance, you can let it stew for half an hour before encouraging it to creep up for a gentle fuss. Should it respond with increased rage, or just excitement that you have invented an even better game, then you have not been anything like firm enough – and the pup is giving you a second chance to do better.

It is unlikely that you will overdo things to the extent that the pup will lose all confidence in you. If this appears to happen you can work hard to restore its confidence with plenty of affection. It will need this and will respond, just as it would in a pack situation. You have become a fearsome 'God' to it and your approval will mean more than anything else.

I have written the above with grave misgivings. We live in an age of political correctness, where any form of violence is condemned and the perpetrator pilloried. Many people, some of them experienced dog folk, will disagree strongly with me. Much depends on the individual animal's personality. Pet breeds are usually selected for their willingness to cohabit amicably with their owners, but this book is about working dogs. The selection of their parents as breeding stock was made primarily for their abilities in the field and, being largely handled by professionals, they never had the chance to display much aggression. It would be quickly crushed. But these same handlers would have valued the initiative that accompanies aggression, and since the latter did not pose a problem for them, would not have been put off breeding from them. So many good working dogs may have the capacity for aggression in their make-up.

Over the years we have changed the temperament of many breeds, to make them more affectionate and easier to control. This has been accompanied by a softening of our approach to dog training, and the development of less repressive methods. As long as owners have bought a dog that has been designed not to challenge our authority or threaten us, then the modern approach is seen to work. But, while some breeds have been changed in the last fifty years, this is not true for all of them. Ironically our security forces now have problems getting suitable dogs to train for criminal-work, as much of the drive has been bred out of them. As a result, many police forces are abandoning their policy of recruiting unwanted pets and find they get better results by producing their own stock from good workers.

The good news is that your puppy will tell you at quite an early age if it is going to be a problem. It will also tell you by its behaviour if your control is sufficient. When you have to chastise a puppy do it quickly and firmly. Then return to being a friend once it has stopped being a nuisance. It will choose which persona it prefers you to have, provided you make the distinction contrasting enough. The results will show if you are making matters clear to it!

My understanding is that dogs make good servants and bad masters, and that their grasp of the subtleties of social hierarchy is infinitely more refined than ours. With pet dogs you may reach a compromise where the dog exists very comfortably in an undemanding environment, because it is only asked to do those things it is happy with. In the world of the working dog the rules are different. Obedience must be unquestioning (the dog lying at my feet has just collapsed in the canine equivalent of hysteria at this concept!) as must be your authority.

Many dogs that have subsequently become unmanageably aggressive with people, first demonstrated this trait as puppies – and got away with it. Do not let your pup think it has the upper hand from the start, or you will both regret it. You do not want to own a dog that is dangerous or out of control. It may injure someone, whether deliberately or by accident, and you will be the person who is held responsible under the law. Neither do you want a dog that you cannot let off the lead – this would only give it a half-life. Of course, it will not be you who is put down if the worse scenario occurs.

Let's move on. Your pride and joy is still wobbling about on unsteady legs, greeting you ecstatically and probably peeing all over the carpet. Will it ever become the definitive working dog you have dreamed about?

BEFORE GETTING A PUPPY

Choose your breed to fit the planned workload and your personal and family circumstances.
Stop if you do not have the time or space for a dog.
Budget the time and money your new dog will cost.

Questions to ask yourself

Have I seen the breed working and met them socially?
Have I located a recommended litter?
Have I seen either or both parents?
Are they free from hereditary health problems?
Is my name on a breeder's waiting list?

Jobs to do

Is my garden securely fenced?
Have I bought an indoor cage, bedding, lead, collar, dog bowl etc.?
Have I identified a good local vet and dog-training club?
Have I made sure that someone can be with the puppy most of the time?

Chapter 3
Laying the Foundations

My late father-in-law, a dog man throughout his life, once said, 'A dog only needs to know four things – to walk to heel, to come when it's called, to sit and stay and to give over.' By this last criterion he meant that he expected a dog to be no trouble. In fact, he demanded very much more of his dogs, without even realising it. Without exception all his dogs were pleasant, willing, obedient and confident. He was able to take them anywhere with him (and usually did), knowing they would not disgrace him in any way. He was gentle with them and they worshipped him in return.

Why is it that some people always have nice dogs? Do they just choose well or do they consciously or unconsciously mould them to fit their expectations? This I think is the secret, so let us try to identify what makes a good dog and then create an overall plan and strategy to achieve this.

Socialisation

Once the puppy is clear of its inoculations you need to take it everywhere with you – it does not have to be hard work!

Firstly, we should understand that to produce a good dog, we have to work on the basic material, the quality of which has been selected in the last chapter. Just as the artist primes the canvas, the sculptor wedges the clay and the cabinetmaker seasons the timber, so we must shape our puppy to have those qualities that we will need when it starts work.

The dog will live within a complex society. It will need to fit in as part of the family 'pack', within which it can operate confidently and comfortably. It will also have to adapt to a wider society of neighbours, passers-by, visitors and so forth. It has not only got to commit to these, but must tolerate and be friendly with them. Its world will also include other dogs and animals, from the family pets to farm livestock and wild creatures. There will be a host of strange things too, such as cars, buses, pneumatic drills, farm machinery etc. A well-socialised dog takes all these in its stride.

Social skills in the family

It is second nature to the dog to understand the hierarchy of its pack, which is now represented by your family. Finding its place within that structure depends on your understanding of this. Any dog worth its salt will aspire to be pack leader. It will, as part of its development, test the boundaries of acceptable behaviour and the strength of the existing alpha animal. In the family environment this means that it will challenge your authority and rules in just the same way as a human adolescent will do.

Very clearly you should be the powerful figure who is the first source of comfort and the ultimate authority, and to whom the greatest loyalty should be accorded. You will have achieved your position, in the dog's eyes, by being prepared to dominate and punish, as well as rewarding good behaviour. You may not actually have to use force, as your body language and tone of voice will have been correctly interpreted by the dog. In a canine society this awesome authority is seldom challenged, although any insubordination will be put down quickly with a fearsome display of aggression until the unfortunate upstart is thoroughly submissive. Having asserted itself the pack leader does not have to continue to beat its followers up. They will automatically accept its authority, and will be very willing to take advantage of its power and strength should danger appear.

In our civilised world we are not used to being despots, and we mostly produce dogs that are unlikely to pose a physical threat to our position. This does not mean that our authority is not questioned. On the contrary our dogs will always push the boundaries. Even old dogs will never stop aspiring to greater privileges, something I discovered when my geriatric, biddable old favourite suddenly decided that he had earned the right to get on the settee. If these small battles are not to be lost at the very start, we must be prepared to be firm and consistent.

There will be lower level 'chiefs' within the family. Some will have more natural authority and status than others. The puppy will see that parents exercise power over the children by giving instructions, rewarding them and telling them off, providing food etc. Equally the children reflect their position by 'begging' from the grown-ups, and from their visibly subservient role in that they have to ask permission to do things or go places. All expect the dog to pay them some attention and will resent the dog not being respectful. Even the smallest child supposes that the dog will have to give the doll back if they tell it!

The puppy will learn most clearly if you do not give it contradictory information or instructions. This means you need to agree as a family what is required of the pup, at whatever age or stage it is at. If you all agree that you do not want it wetting in the house, then you should also agree that it does not matter who sees it wake up – as long as they rush it outside and praise it when it performs. It is everybody's responsibility and the pup will learn that everybody has a right to direct it.

You will want the puppy to like the whole family, so it is agreed that you can all play with it and take turns to feed it. All the family will have the right to take things away from the pup. The kids will rescue toys and shoes, Mum will snatch back clothing stolen from the laundry basket, and you may need to prise its mouth open to retrieve dangerous, toxic or valuable objects. The pup is learning that it is subject to the dictates and will of you all, which is as it should be.

Even at this early stage you can start avoiding problems. Agree that the pup is not given food at the table, as you do not want it begging and being a nuisance. This might mean you agree to put it in its crate at mealtimes, which any member of the family can do. Agree that when it jumps up, anybody can slap it down. A sharp tap on the nose and a cross 'No!' will soon dissuade it. Anybody can crossly get it off the furniture, so that it learns that the floor is the proper place for a dog. How nice then that visitors can come in without the dog jumping all over them, or that it does not leave muddy footprints on the settee in wet weather.

These are simple rules of good behaviour, but it is surprising how many people indulge a puppy and allow bad habits to creep in. Later on, when they become a problem, you may need to be firmer in changing a learnt pattern of behaviour and your dog will be confused by the reversal of the rules. So, agree the ground rules and keep to them.

Aggression

It seems silly to tackle aggression when we are still dealing with a tiny puppy. However, it makes sense to do it now, rather than when it is bigger and may represent a genuine threat. Within the litter your pup will have fought with its siblings, and anyone who has reared a litter of pups will know how fierce their small confrontations can get. I would be prepared to bet that every dog that shows aggression in later life got away with it as a pup.

Within your family, its new pack, the puppy will still play-fight and test its strength. This is a necessary stage in its development, and sharing in its games will create a tremendous bond between you. But you can still choose to make an issue of some sorts of bad behaviour. Do not let it growl over food or toys. Do not let it complain when you pick it up to put it to bed, or move it while it is sleeping. Its brothers and sisters will have walked all over it without it objecting, so it should show you the same tolerance.

If the puppy does growl, and even tiny pups can demonstrate real anger, you must act promptly. Grab it sharply by the scruff of its neck and really snarl at it. Of course, it will be upset (and in the unlikely event that it is not, you may need to keep a close eye on any future signs of aggression) but then you can comfort it and reassure it. Let it know that you do not dislike it – only its behaviour.

If aggression persists, and it very rarely will if stamped out from the first, you will need to get help. There are many good trainers and behavioural therapists around. Find one in your locality quickly. Never laugh off infant tantrums or, worse still, encourage them. You want a pointing dog not a pit bull.

Introducing the collar and lead

As the puppy settles in you will want it to meet a wider circle of acquaintances. Before this it must accept a collar and learn to walk on the lead. An adjustable nylon collar is fine at this stage, as it will soon become too small. Put the little collar on when the pup is due for a feed, so that it is distracted by pleasant things. At first it will scratch to remove it, but playing with the pup will take its mind off this strange 'instrument of torture'.

After a few days the collar is more or less forgotten. Then, in the comfort of your own sitting room or garden, introduce the lead. Suddenly the collar is remembered again, and the puppy struggles to escape or back out of it. Do not put any pressure on the lead but keep it slack – apart from having to disentangle the pup at intervals. Then encourage and coax the pup to follow you or come to you. With several false starts it will gradually get the hang of things and will accept the restriction of the collar and lead.

A normal reaction to a new collar

Some pups do violent head over heels, lie down in a complete strike or try to back out of the collar furiously. Carry on, dragging the pup gently along if necessary, with lots of praise for every step it takes. By the time it is ready to go for walks it should be fairly relaxed about the whole enterprise. You will also have had the chance to ***make sure the puppy cannot get out of the collar,*** a very important consideration before you go on to the streets.

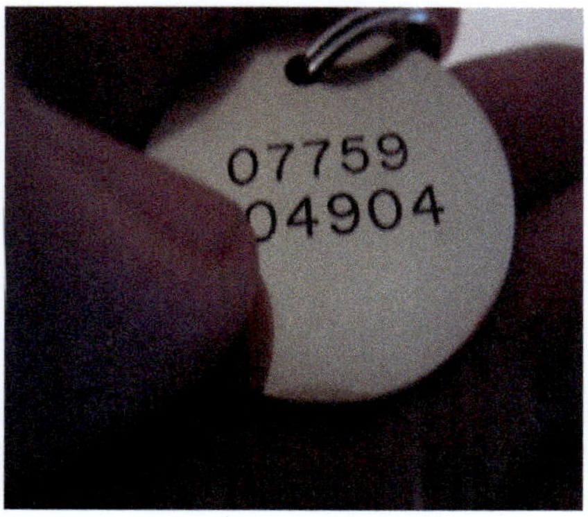

Before free-running your pup, remember an engraved address tag is a legal requirement. Only have your telephone number put on this. There is nothing more annoying than well-meaning strangers grabbing your dog's collar to learn its name, and then insisting on trying to get it to respond to them. It is also unwise to put your address on the tag as there is a danger that thieves may be able to trace it to your home. There is a very active market in pedigree puppies and you do not want your pet to become part of this trade!

NB Microchipping is now a legal requirement and your puppy's breeder should provide the certificate of this at the time of purchase – but do make sure you register it in your name!

The wider society

This begins with the vet, when you take the puppy to have its inoculations. The vet will be gentle with the pup because he or she will not want to put it off coming into the surgery again. But your vet will also be firm, so that the puppy learns to accept being handled by strangers. When you think of it the puppy will not be physically restrained by an unknown person for some time to come. Practically everybody else you meet will either make a great fuss of it or will ignore it.

Although I am not infatuated with showing my dogs, they all attend the local ring-craft class as soon as they are clear of their inoculations at twelve weeks old. They are given the chance to socialise with other dogs, and are routinely examined and handled by total strangers, who do not try to make them do anything. It is not as noisy or full of disobedient dogs as a training class might be. Also, I want to plan my own agenda of training and do not necessarily want to be in a training straight jacket, designed to comply with someone else's timetable.

Your puppy will also be going out for walks on the lead, preferably ending up at a safe area where it can be let free to explore. At this stage it will lack confidence and may bolt for home, perhaps having or causing an accident. Make sure it cannot easily get on a road. However, remember that you are the most familiar person to it, so it will be very hesitant of straying too far.

A safe area allows the pup to develop its confidence, and you can let it enjoy its freedom.

Getting it back should not be a problem in its dependent state of mind, but you can still exploit the situation. Introduce the command 'Come' (or the appropriate whistle signal) along with its name, whenever it comes up to be fussed. Habituating it to the intended sound in any given situation is always a good idea. At this age it will associate the noise with the pleasure of your praise and attention, and not resent it because it is the end of its liberty. Also get it used to running after you for play, or for chasing a toy, so that your company is always seen as being the best thing.

Expect to meet other people while exercising your pup. Dog walkers form a kind of Freemasonry and are usually happy to fuss over a puppy when it runs up to them. If their own dog is likely to play roughly or even be aggressive, they will normally try to get it under control before an incident occurs. If you meet a friendly, kind dog, and its owner has time, take advantage of this canine and human goodwill to let your puppy have a good game. After all it is not long since it had brothers and sisters to play with, and the memory will come back quickly enough. Your pup will look forward to going to the park tomorrow, and walking on the lead will soon become a treat and not a torture.

Not everyone will want to be jumped upon by an enthusiastic puppy. Joggers are concentrating on running and do not want to trip over your pet, nor do you want it to run off after them. Jumping up at mountain-bikers is also not recommended for either dog or cyclist. Other people, going for a walk in the country, often wear quite unsuitable clothing, and get annoyed when your puppy signs its autograph in mud on their pale trousers or coat. There are also an increasing number of overprotective parents who unwittingly teach their children to be frightened of dogs. These children scream and run away, which is an open invitation to your 'ferocious' hound to go careering after them. If it then jumps up and catches its baby teeth in their garments, be prepared for an ear-bashing at least. Pretty soon you will need to stop it jumping up.

The socialisation process should continue throughout your dog's life. The more relaxed it becomes in an increasingly noisy and busy world, the more reliable it will be when working. The sophisticated 'dog about town' will not be as upset by the unexpected as the one which has led a sheltered life. Think of how socially integrated a guide dog for the blind needs to be, and copy the lengths that very successful organisation takes to achieve this.

Farm animals

Dogs that work in rural areas also need to be familiar with farm animals. The countryside has its own smells and sounds. Tractors can be very intimidating and noisy. Cattle are inquisitive and may run up to a young dog. Sheep often run in the opposite direction, which is another temptation your pup will have to overcome. It will need to realise that they are not some kind of woolly dogs to be played with. Poultry can equally become victims.

A dog that cannot be trusted with livestock will never be welcome in the countryside, so you must develop its social skills before it is big enough to do any damage. Since we intend to shoot or hawk over the dog, I assume that access to farmland is not a problem, as this is a prerequisite of both these sports. The farmer will be more than eager to help ensure that your dog will not represent a danger to his stock, and will almost certainly allow you controlled contact with them.

The ideal response to farm animals, that you are trying to develop in your dog, is indifference. Too severe an introduction can make it reluctant to enter stocked fields, which is no use if fallen game has landed there. Your dog may also need to quarter moorland where sheep proliferate. If the dog becomes too wary of cattle, and clings to your side, this can put you at risk as cattle are known to sometimes attack people with dogs at heel.

I was once put in hospital in just such circumstances on a regular walk across a farm holding a herd of beef cattle. For some reason I called the dog to heel to avoid disturbing them. One individual cow, with a yearling calf, followed me, broke into a trot and knocked me to the ground, where she commenced trampling me. Luckily, I was not alone and the efforts of my companion (aided by the fact that the dog had heroically run away) stopped the attack before I was too badly injured. Subsequently I learned just how frequent this kind of attack is.

Walk the pup along lanes flanking stocked fields. Let the puppy sniff at cattle, horses and sheep through the fence. It will probably be nervous of these giants as they huff and puff at it. Do not be overly sympathetic. Sometimes dogs learn to exploit this and, increasingly, may play on its sensitivity to get attention. You must be casual, relaxed and a bit dismissive of its fears. It will see your lack of concern and take its lead from that. Neither should you try to make the presence of livestock exciting. You could encourage it to want to play.

Nothing to get excited about

Once the puppy has some experience of other animals you can seek out more demanding situations. I like to take a puppy through the lambing sheds. You can bet some of the older ewes will be more than willing to present a hostile face to it. Or, if the time of year is against you, take it to a livestock market and stand next to where the sheep are being unloaded and driven into the pens. There will be a lot of clattering and banging, dogs barking and men shouting. The bustle will be both exciting and frightening. Control any excitement your puppy shows with a few stern words and a light slap of the lead on the pup's bottom, or an equally light jerk on its lead.

Similar experience can be gained at your local farm when they are gathering sheep for clipping or dipping (and giving the farmer a chance to see your pup behaving in these circumstances will not hurt your access to his or her land either!).

Poultry can usually be introduced in small pens where you can plan the meeting with the pup on the lead. Geese can be relied on to scare pups – it is always a good idea to try and get the animal or bird to provide the aversion therapy rather than you.

Litter mates but the dog (on the right) lives among sheep and has got completely used to them. His sister is more likely to investigate.

The important thing is to take the time to plan introductions to livestock while the puppy is still vulnerable and unlikely to be assertive in its response. ***Waiting until the situation arises is stupid.*** You may find your pup has run after sheep in play, but you cannot get it back, or it has killed a chicken. These may be hard habits to break, and can lead to a collapse in your control.

Birds of prey

Most keen falconers intend to use their pointing dog with their birds, and this combination forms an unholy alliance that is mutually beneficial. A young puppy can be introduced to experienced birds while they are tethered on the lawn. The pup's approach will be a tentative one as this new creature has an unusual smell about it. I do not bother to put the pup on the lead for this lesson.

A bird that is used to dogs will permit the pup to get close to it – and may then have a dive towards it. Probably this will be enough to warn it off, but bolder pups may interpret this as a sign they have found a new playmate. Relax and let nature take its course. Most good hawking dogs have learned the hard way that talons hurt. This is no more than the puppy would experience from older dogs in the pack situation, so it is able to cope with it. Of course, if your bird is young itself, or small enough to be vulnerable to a puppy, you should supervise this introduction on the lead and be prepared to be very firm with the pup if it oversteps the mark.

This Brittany pup, at six months old, is already very relaxed with the young sparrowhawk that will be its future hunting partner.

Taking your bird with you, while you are taking your pup for its free exercise, is good practice. Both will learn to associate the presence of the other with pleasant experiences, and the puppy will have other distractions to take its mind off the bird. Once they begin working together, they will develop a great affinity and will be surprisingly tolerant of each other. I have seen my hawks tweaking hairs from the back of my old dog while he lay beside their perches, and then letting him lick their feet after they have finished eating. He would not dare to interfere with them on a kill, but they are more than happy if he helps pin down large quarry that they are struggling to overpower. When hunting I find that both dog and hawk cut out the middleman – and work with each other while ignoring me!

Mixed terrain

A good dog should work in any kind of country. It makes sense to slowly build their confidence by starting with easy ground and gradually finding deeper and heavier cover. Avoid nettles, thistles and brambles, as these are calculated to put a pup off trying to push through dense, tangled cover when they start working in earnest. It is sufficient to just challenge their eyeline – so that they get used to losing sight of you while they investigate. Always make it easy to start with, and build on this success. Once it learns to retrieve, you will be able to challenge its nose and initiative by throwing a favourite toy or dummy into easy cover – and progress from there.

In open ground, i.e. where there is good visibility and little cover, your pup can see you easily. This will give confidence to a clingy pup and will encourage it to range more freely – which is good. Do not be in a hurry to enter woodland or deeper cover, where it might easily lose sight of you. Conversely, with bolder individuals there is much to be gained by deliberately trying to hide from them. By keeping them insecure you will teach them to always keep one eye on you for fear of getting lost. This should be a continual exercise at this age. The pup that continually does this will never be difficult to get back. It will keep coming to you for reassurance and praise, which is better than continually having to nag at it to come back.

Start with light cover and slowly introduce more varied, but still inviting, cover as it progresses.

Early water training

Treat water as just another form of cover, which in a way it is. Make early introductions as easy as getting your puppy to walk through puddles, before moving on to wading in shallow streams. You will eventually want it to be sufficiently confident to plunge into fast-moving rivers, or to jump into water from a high bank. Being too ambitious too soon can have the opposite effect and, if you 'over-face' your puppy, it may develop a fear of water instead. Make haste slowly – lots of splashing about in the shallows will lead to familiarity when it is old enough not to need 'water wings'. Much better for any pup to come to water when it is strong and confident, than to have it struggling too soon.

These two youngsters are much too interested in investigating new things to be afraid. For them water will always be something they just cope with.

Negotiating fences

Another necessary skill for your dog will be to negotiate fences, gates, walls, stiles and any number of different obstructions. Although jumping must wait a few months, until your pup is agile and strong, you can teach it to 'Get under'. This is done by coaxing it under wire, rails or low branches so that it learns to follow your hand. Apart from giving it the confidence to tackle hazards, you are also introducing signals that will be part of your directional commands in the future.

With the incentive to get to the other side it will soon associate this hand signal with squirming under things.

I tend to use 'Get through' if I want my dog to pass through a stile or fence, and teach it in exactly the same way. When I eventually add 'Get over' for jumping and 'Get in' for entering cover, I find the dogs pick this up very quickly.

Jumping can damage a young dog's growing joints and should be left until your pup is virtually an adult. You will need to make a subjective judgement about when you can introduce it. This will vary from breed to breed according to their basic agility and also depends on the maturity of the individual. Very obviously scrambling over a log is very different from clearing a barbed-wire fence, and small progress can be made with easy obstacles from an early age.

An ideal situation. The undergrowth leads the pup into the only gap and there is no easy route around or over.

Use 'Under' or 'Get under' rather than calling it through the gap – here the exercise is to tackle the obstruction, not practice recall, and you cannot always be the other side of an obstruction. Once the pup has become confident at reaching you, by squirming under the gate, it will readily crawl underneath with you beside or behind it.

FUNDAMENTAL CONCEPTS

One of the core dilemmas of dog training is achieving the correct balance between initiative and obedience.

Both these qualities must exist in a pointing dog but they are contradictory. Concentrating too much on obedience risks eroding the pup's initiative and its decision-making potential. Equally, allowing it to develop too much self-confidence can lead to serious control problems in the field. Getting the right balance is vital to its working life. Because you can only guess at the olfactory landscape within which it lives, you are totally dependent on it wanting to find game. At first it appears that your pup is running about mindlessly instead of focusing on its work. Contrary to the visual evidence, this is unlikely to be the case, because you have the benefit of countless generations of breeding on your side. If your dog is generally obedient you can be more relaxed in that, since it does not normally mess about, it is not doing so now.

Initiative – the ability to act independently

You will be only too aware that any young dog is usually very willing to act on its own initiative – and much of its social training is planned to ensure that this does not happen. It should not urinate wherever it pleases, nor chew and destroy your possessions and furniture. Neither should it reduce your garden to something resembling a battlefield. Importantly, you do not want to make it so anxious to avoid punishment that it is incapable of doing anything without your permission. Pointing dogs are there to react to scents that you cannot detect, and they must feel free to make their own decisions. However, you hope that, in trying to please you and get a reward, they will use their abilities to achieve what you want. But without sharing their scenting ability, you have to trust them to sort it out for themselves.

Look at the various components of the pointing dog's portfolio of skills and you will see that they are mostly based on things the dog wants to do naturally – hunting, pointing and retrieving. Try to develop all these by encouraging the dog that these are fun activities that you approve of. In consequence it will approach them confidently in the knowledge that they are desirable.

Your training will be based on positives, and on gradually setting more challenging and complicated puzzles for the dog to solve. As it starts to apply itself mentally to the game it will get more confident and fulfilled – provided you have increased the degree of difficulty so slowly that it is never over-faced and disheartened. With each success its confidence and self-belief will increase, as will its determination to triumph.

Using the retrieve as an example, you would start with the simplest play, with your puppy on the sitting-room floor. Then move on to finding toys in very light cover, before using dummies, cold game and, eventually, freshly shot birds in the deepest of cover, or over water and fences. The experienced dog approaches each new situation with confidence and enthusiasm, because it has learned that you always send it for attainable objects. It will find ways around or through obstacles, and will learn how to track runners or mark unattainable game.

You cannot make your dog find and retrieve game – but you should make it want to!

Obedience – responding to instructions without question

Achieving obedience is a different matter. It is based on two assumptions – firstly that you know best and, secondly, that the dog will accept your authority to make decisions. The first is not true in scenting situations, so you cannot overrule the dog's knowledge (although you may need to overrule its actions if they are dangerous, e.g. wanting to pursue a runner across a busy road). Whether your attempts to stop your dog are successful will depend on its respect for you, even in the face of great temptation.

I can do nothing about the accuracy of your judgement – that is something you will learn or not from experience. However, your authority needs to be developed from quite early in your relationship with your dog.

Is authority automatic?

Emphatically 'No'. It has to be earned and exercised continually. Dogs understand hierarchies as well as we do, and exploit any sign of weakness as ruthlessly as your children will. Every pack has its alpha male and female, and they do not get there by being conciliatory! They will retain their position jealously and at all times. The rest of the pack enjoy a protected and comfortable life by accepting this structure, but some of them (those with the greater initiative) will be prepared to challenge authority at the first sign of weakness. There is a progression up the pecking order as pups become juveniles and then mature animals, before themselves showing vulnerability in old age and slowing down. In the domestic dog the family is the dog's pack, and its position on this social ladder will alter as it grows and matures.

Most challenges to your authority start during puppyhood. The puppy that chews socks or shoes, and then runs away with them to avoid their removal, is evading authority. Its failure to come in promptly from the garden when called is another good indicator, as might be its insistence on climbing on the furniture. It is not consciously setting out to undermine you, but it is learning that it can do what it wants – which will have the same end result. The sensible dog trainer will identify achievable ground rules from the start, and will make sure that the puppy is not allowed to get away with anything – although you will not overly correct bad behaviour at this age. It is simply a case of having your way in a kindly but determined fashion.

In the next chapter we cover obedience much more fully, and this book assumes that you will need to continually refer from one section to another throughout your dog's education. ***Obedience becomes serious when you need to insist on the dog responding properly, even when it does not want to.***

Timing

Correct timing is vital to the pup's understanding of you. Being effective works only if the command or signal, and the resulting contrast of praise or correction, is so quick it can easily be identified with the situation we are reacting to. ***Your correction must be instant and effective. Equally your praise, as it responds correctly, must be immediate and fulsome, especially with the young puppy.*** Older dogs will understand why you are displeased with something they have done some time before (do not listen to those experts who tell you dogs do not have a sense of guilt), but a puppy cannot reason like this.

The dog's mind is working all the time, just like ours. To help it make an association between your commands and the required actions, you have to be extremely quick and accurate. Sometimes this amounts to actually anticipating the dog's action, so that your command coincides with it performing the action you want. For example, when teaching the retrieve, you can throw a favourite toy for the puppy with a reasonable expectation that it will run after it. The moment it reaches the toy, and as it bends down to mouth it, is exactly the moment to say 'Fetch'. Equally, as it grasps the toy (one millisecond later) you need to praise with a delighted 'Good dog, fetch'. To the onlooker it sounds like 'Fetch, good dog fetch,' but the dog will associate the sounds specifically with what is going through its brain at each split second in time.

When you are not concentrating sufficiently on what the dog is doing, it is easy to give conflicting messages. Saying 'Fetch' as you send the puppy out can lead you (and it) into confusion, if it then ignores the toy for an interesting smell. It may interpret 'Fetch' as meaning 'Go out and play', which is not what you intended at all. You can further confuse matters if, after picking the toy up, it then drops it to investigate that smell again, just as you praise it for carrying the toy. Your 'Good dog, fetch' then becomes 'Well done for dropping the toy and having a sniff'. By being slightly slow in your speech and attentiveness, you may actually be teaching your dog not to retrieve!

For people who are not quick enough, there is something to be said for learning clicker training as a means of accurately synchronising the correct timing of your response to the dog's desired action. Ideally you would have to be pretty skilled at using a clicker before training your puppy, but it would be difficult to learn this without a dog. You will need experienced help to master this. Just get used to being alert, and concentrating on your puppy while you are training it.

Being effective

Apart from poor timing, the most common handling fault occurs when the handler fails to be effective. Whenever you find that your training is not going well, or the dog seems particularly dense, we can usually trace it back to this simple failure. Accept that you cannot reason with your puppy or explain what you want it to do. Instead make the required action so desirable that it will feel excited at the prospect of pleasing you. To do this use verbal and physical praise, and sometimes food rewards. Fortunately, the actions we demand of our dogs are usually the ones that they really enjoy – so training them is self-rewarding.

Alternatively, when your dog does something which is unacceptable or undesirable, you may have to chastise or punish it. Verbal reward does not work if you simply tell the dog, lucidly and rationally, why it is a good idea for it to listen to you, process what you have said, and then carry out your instructions. Equally a lengthy and reasoned explanation of the possibly unpleasant and negative consequences of bad or unsafe behaviour will not meet with any response. The dog has no understanding of complicated sentences or words, and learns only sounds.

Fortunately, dogs have a spoken language of their own. All dog owners can differentiate between the yelp of a frightened puppy and the menacing growl of a seriously nasty guard dog. We can recognise the plaintive whine of a puppy shut away from us, and the playful bark of one that is challenging the sprinkler on the lawn. Dogs are surprisingly vocal without resorting to words and, instinctively, we understand their language pretty well. They also use tone and volume. The louder they make any noise will reflect the strength of the message they are sending. In just such a way they will understand the tone and volume of our voice.

Dogs also have a second, complex language which consists of posture or bodily stance, facial expressions and gestures. Such clues as ear and tail carriage are familiar to us, and we can understand the stiff-legged walk and arched neck of two equal dogs checking each other out. Any of us can separate the meaning of a puppy barking at the sprinkler, and one barking at a strange noise. In the first instance it is accompanied by splaying its front legs apart, rump in the air and with its tail wagging vigorously. The frightened bark is seen with raised hackles and the body poised for flight.

What all this means is that you can speak to your dog through this common language. By using your own posture and tone of voice you can display happiness or anger. Dogs will understand these, but will also read telltale signs if you lie to them. Saying 'Good dog' in a cross tone will not convince them you mean it. So how do you check that your message has been understood? It's easy – your dog will tell you.

Being effective is simply a matter of judging by results. If the dog responds with greater willingness and confidence, the next time you give the command, then your praise has been effective. If it shows less confidence, or is more reluctant to respond, then you have failed to effectively praise it, or have given a contradictory message.

Very clearly this posture tells you the dog is being playful

In just the same way, when correction or punishment is required, you should see an improvement next time – **if you have been effective.** For example, if your puppy chases poultry it may be necessary to punish it verbally or physically. If, on the next day, it goes out and tries to commit the same crime, then it is telling you that it was not very bothered by your discipline. In contrast, should it hold back and try to avoid the chickens (which it now associates with your displeasure), you can claim to have successfully applied aversion therapy. You are then in the position of being able to praise it for ignoring the poultry, and have turned a negative issue into a positive achievement for the puppy.

The ability to be effective depends very much on your willingness to be exuberant with your praise or fearsome in your displeasure. By creating a great contrast between the two extremes you will be able to explain to it the course of behaviour which is to be most rewarded and desirable. This black-and-white explanation is kind to the dog, as it helps it understand us clearly. If you are self-conscious (and most British people are desperately so) you will probably only deal in 'shades of grey'. This lack of contrast makes the dog's task of understanding you very much harder.

Quite a few HPRs will overreact to both praise and correction, by either becoming excited lunatics or squealing like stuck pigs. However, **the test of effectiveness will be the dog's behaviour the next time the same situation arises.** If you have not got through to it then be prepared to ignore their overreaction or, better still, treat it as another training opportunity. You should be able to praise or correct your dog without it using this as an excuse to avoid obeying. For example, it is perfectly in order to correct an overexcited dog for behaving like a fool, or to tell it off further if it feigns terror or agony at some minor punishment. Evasions occur at other times, and in all sorts of ways, but over-reaction to control is an example.

The famous television dog trainer, Barbara Woodhouse, said that she got into dog training because she had a rescue Great Dane bitch that was extremely nervous and neurotic. She went to great lengths to create positive associations with these fears, for example feeding the dog on railway stations while trains came through. After prolonged efforts with a complete lack of progress, she became so exasperated that she lost her composure, and corrected the dog for being such a stupid fool. To her surprise the dog immediately shrugged off its fears, and from then on became perfectly indifferent to them. Barbara realised that her sympathy had actually been rewarding and therefore encouraging its negativity.

Responsiveness

Responsiveness is something to really value in dogs, but it is not a constant and can be reduced or increased. It is the sum of the dog's willingness, added to its sensitivity to both sound and touch. All these qualities can be altered for better or worse by effective handling. We know that a human boxer, as a result of continual bruising encounters, is likely to become impervious to pain that would really hurt the rest of us. Similarly, a teenager, spending a lot of time listening to loud music, is prone to hearing loss through this continual exposure. If either had this continual bombardment removed, they would soon react to a lesser level of stimulation. In practical terms sensitivity measures the reaction to anything that is in contrast to the norm, so if the normal level is low the contrast is potentially greater.

Sensitivity to sound

We can choose how much sound to bombard our dogs with. If you think your potential shooting companion is quite sensitive to sound it would be worthwhile subjecting it to constant high levels of noise – particularly in association with pleasure. Clatter its food bowl about when feeding it, and speak in a loud voice when giving praise, even clapping your hands in delight at its successes.

With a less sensitive dog you should minimise background noise so that quieter commands will stand out in contrast.

Sensitivity to touch

Similarly, physical conditioning can be used with timid pups that may be reluctant to enter cover or water. Try to establish lots of rough-and-tumble games, and give boisterous pats and heavy stroking when praising it. It will toughen up slowly, ***provided you persevere, and never go overboard to the point where your praise or affection is seen as a punishment***.

Willingness

This quality is variable as well. By increasing it you will add to your dog's responsiveness – for better or worse. A sensitive dog that becomes excessively willing, will verge on a state of continual anxiety, for fear of making mistakes. With HPRs this is often associated with false pointing – when the dog tries to turn every possible scent into quarry. Praising it less energetically will reduce the self-imposed pressure it is under, and make its life much more endurable. Be patient, make your voice reassuring and your whole attitude unhurried. Give it time and confidence to overcome its fears. Developing its self-confidence in this way will help with its initiative and, with such a sensitive animal, obedience is not an issue anyway.

Where a dog is physically quite tough you can increase its responsiveness by building up its willingness. The lazy approach, of just nagging at it with continual corrections, does not help. In fact, it will probably just make it tougher. It may become a thug and will tend to self-hunt and chase game. **However, even the toughest dog can be a responsive and bold worker, as long as you can create and maintain a high degree of willingness.**

Of course, we all have our own natural level of handling and the dog learns to adjust to it. But if you feel that the amount of effort you are putting in is not achieving what you want, you might look at the dog's sensitivity and willingness levels. Do not blame the dog. It is what you have made it.

Early lessons

There is no set speed for your puppy to learn at, nor an exact syllabus it should follow. Each puppy has its own pace, determined by its confidence, your ability to teach it, and the opportunities you can give it. Instead of sticking to a predetermined timetable, look at your pup's overall development. Work on exercises that will develop its skills and personality as you would like. As long as you are clear in your own mind of how you want your puppy to behave in the future, you can avoid it doing those things that are contrary to those plans.

With a pup that is sensitive, clings a bit too much, or is frankly 'windy', try concentrating on aspects of its work which will develop its initiative. Think of lots of free-running in the local countryside to familiarise it with the smells and sounds. Interest it in game smells, such as occupied rabbit warrens, and play with retrieve items that it likes. You can introduce negotiating fences and gates, or splashing about in water. Do not place emphasis on heelwork, sitting and staying or recall. Give lots of robust praise to toughen it up.

A bolder pup, with the first regime, would quickly become a wayward and overconfident hooligan. Instead introduce basic obedience exercises in a fairly gentle, unambitious way, but insist that it ends up doing what you wanted in the first place. In outdoor areas still demand that it pays attention to you before it gets its freedom. Look for little battles to win. Although this might seem laborious progress a puppy like this will soon catch up once it has to use its initiative.

Chapter 4
Obedience

Obedience is very much the cornerstone of HPR training. It requires the dog to respond to your commands and signals, even when it does not want to. This means that it respects your authority, and is anxious to please you and earn your praise. If you have any ambition of working an HPR successfully you must really achieve this sort of control. Otherwise it will find a thousand opportunities to disobey, such as chasing rabbits, 'pegging' game or eating carcasses instead of retrieving them. These all sound pretty fun activities for the average dog, and it could easily add many more to its repertoire.

There are some very basic rules about obedience. Firstly, it is not something you switch on and off. Obedience is a permanent state of being for the dog, and you should always demand it. For the trainer of a house pet this imposes a discipline. He or she must ensure the dog is never allowed to get away with anything that has been previously forbidden, and must insist that all commands are obeyed. This is difficult to achieve, and is one of the reasons professional gundog trainers kennel their dogs. They only interact with them for work or training. By contrast, the owner of the house dog can develop a closer bond and understanding than these professional trainers need.

1. **Have all the family use the same set of commands**
2. **Only give a command if you are in a position to enforce it.**
3. **Do not give a command unless you are willing to insist it is obeyed.**
4. **If you have to repeat a command it has already become optional.**
5. **Do not repeat commands endlessly, you are teaching your dog to ignore and disobey you.**

Start by getting its attention

Before giving a command, make sure the puppy is listening to you by calling its name first. You will already be doing this in all sorts of social situations, without necessarily associating it with any actions. Soon you will be able to talk to it, without it interpreting its name as a signal to move. Until you have its attention it is pointless giving any command – as you will just be dooming yourself to failure.

Obedience starts in your own mind

- Firstly, you must believe that you have the right, as the dog's pack leader – and the person who incidentally feeds, exercises and cares for it – to expect it to do as you ask.
- Secondly you must be clear what you expect (or hope) the dog to do for you.
- You must select a single, recognisable command or signal for each situation that the dog is to understand and respond to.
- The dog needs to be taught to respond to this signal.
- Successful obedience is the result of your perseverance and determination to have your own way.
- Failure to enforce your authority will only serve to undermine it.
- Failure to reward the dog's response adequately will demotivate it.

Be clear what you want each command (and/or signal) to mean

Most commands will be verbal but, because your HPR will often be working at a distance upwind of you, your voice will not necessarily carry to it. For this reason, some commands also need to have a whistle, and/or a hand signal, to carry the same meaning. Early lessons are done under controlled conditions where the dog is near you and can hear you – so teach the verbal command first. Only introduce the whistle and/or hand signal subsequently, creating an association in the dog's mind between these and the verbal command.

Each command needs to be associated with a combination of small actions, which add up to the intended task. The dog is perfectly able to understand this, and will not need separate commands for the individual elements. For example, the command 'Come' actually means 'Leave what you are doing – return directly to me – stop when you arrive – and sit beside me.' You could teach these as 'Leave, come, sit' but there would be no purpose. If it only receives a reward of praise or food when it completes the whole exercise, it will expect to complete the whole exercise to be rewarded.

The fewer commands you teach the more readily the dog will learn them. They must all be identifiably different, and given in a firm enough tone to make sure the puppy hears them, even if it does not appear to have listened. Do not speak louder than you need – it is always worth keeping something in reserve.

Sit

I have chosen this as a starting point, for the simple reason that it should be the HPR's default setting in the field. You will eventually want it to sit on flushing game, and on shots being fired. In these situations, 'Sit' implies assuming, *and maintaining,* the sitting position – not just the temporary action of sitting. Therefore, from the outset, encourage the understanding that sitting is a continuing state. If the puppy gets up again immediately we need to repeat 'Sit' – not introduce a secondary command such as 'Stay' or 'Wait'. These are superfluous as a dog that is sitting is, by definition, not going anywhere. Be clear in your own mind what you want the dog to do. It will save you time.

Introducing the command

All puppies tend to sit quite readily of their own accord. You want to form an association between the action and the word by repeating 'Sit' whenever the puppy chooses to do so. This must always be reinforced with vocal and physical praise.

Next choose a moment when the puppy is relaxed and in contact with you. While talking to it generally, to keep its interest, you should very gently push its bottom to the floor while holding its head up. The clearly spoken command accompanies the action, followed by more specific praise. Always include the command within your verbal praise, 'good dog; **Sit**; clever dog', with the command being emphasised. Probably the puppy will respond by jumping up excitedly. Wait until the first excess of energy has died down and repeat the exercise. Do not expect instant and reliable understanding. You are just trying to create good habits.

As the puppy becomes more familiar with the sequence of events it will start to anticipate the praise, and sit more readily. Excessive reward from you will cause it to get excited again, so modify your praise to be less stimulating and more reassuring. Introduce physical praise by stroking it gently, so that sitting becomes an enjoyable activity and not an unwelcome interruption. The puppy will come to learn that 'Sit' implies a position, not just an action, and that sitting is pleasurable.

Begin making the pup sit before you feed it, so that it needs to respond more quickly in its haste to get its meal. Given an incentive like this it will sit very promptly on command. However, it has only learned a trick. True obedience starts when we ask it to respond when it has other things distracting it. This may involve telling it off, or physically punishing it for not complying. I can hear a shiver of horror from all the politically correct animal lovers. Ignore them. When you need to stop a dog running into traffic, or chasing livestock, there will be no room for niceties. Commands are not requests and should be non-negotiable and obeyed instantly.

Imagine a situation where your puppy, which willingly sits when it suits it, is being a nuisance. Perhaps it is trying to jump all over a neighbour's small child. It is quite reasonable to tell it to sit but also quite possible that it will ignore you. Having given a command, you now have to be prepared to enforce it. The puppy is intent on having a good game and you have nailed your flag to the mast by telling it to sit. What are you going to correct it for, and how? You have two problems to address. Firstly, the pup is too excitable, secondly it has disobeyed your command.

Your first job is to get the pup's attention and get it under some form of restraint. If it is on the lead you can simply retreat a pace or two, and the inconvenience to others is minimised immediately. Also, the dog cannot get to play, and will be more likely to notice you. If the pup is loose, your problem is more urgent, because it may knock the child down, tear clothing and cause panic. With luck you can call it to you and get hold of its collar, which puts you back in control of the situation.

This is not always easy to do. Faced with an excited pup you may need to get its attention by a fierce tone of voice. At a young age they are still programmed by nature to listen for a sharp, warning 'bark' – so give one. Almost certainly your pup will look round with concern. In a split second change your tone of voice to one of pleasure, and try to get it to come to you. Immediate praise will help it to approach you confidently, and give you the chance to catch it. **Timing is everything**.

Now is the time to tell it to sit again, when you are able to make it obey. Hopefully you will have stopped the puppy being a nuisance, and given it the chance to earn praise by doing as it is told. You will have done all this without antagonising your neighbour, and without getting a reputation as a cruel dog-beater.

The initial problem remains. You must set up similar circumstances but this time with the pup on the lead. Get your own family or friends to play fairly quietly, with the intention of distracting the pup, and then tell it 'No' firmly. Failure to respond should lead to a sharp tug on the lead, a shake by the scruff or a light smack on its bottom. If the chastisement is sufficient the pup will stop momentarily and you should tell it to 'Sit', followed by immediate praise. Keep repeating the exercise until the pup quickly leaves the distraction of play, and sits when you tell it.

Two things can go wrong with doing this as an artificial exercise. Either the degree of distraction offered may be too great for the puppy to ignore, or your control is inadequate. Start by ensuring the pup is not over-tempted, so your task, in competing with the distraction, is easier. Build on success until it becomes familiar with being controlled, and to experiencing a positive outcome when it behaves.

Where your control has been too gentle the pup will not improve, so you will need to step things up a level, becoming sterner in your control and more fulsome in your praise. Its response will tell you how you are doing.

As it gets better you can get your 'decoys' to be even more ambitious in their efforts to get its attention, until you can be fairly certain your pup is controllable in public. You can take this a step further by asking them to gently try to distract the pup while it is sitting. Here you can start telling it off as soon as it moves, repeating the command 'Sit' – because that is what it is now disobeying, and reverting immediately to praise as it sits down again. Done well this can become quite a game and does teach the puppy to listen to you in all situations. Using friends and family is a very good idea, especially if you need to correct your puppy in any way.

When the puppy responds well to your spoken command, you can introduce hand signals. At a time when the pup is being sensible and responsive, perhaps after a couple of minutes of sitting on command, try getting its attention while it is a metre or so away from you. Quietly use its name to attract it and then tell it to 'Sit' at the same time as you raise your hand up above your head, with the palm facing the pup. Think of a traffic police officer stopping an oncoming car.

There is a fair chance it will do this readily, upon which you should try to move back towards it. This should be done calmly but quickly, so that it does not get up. Lots of praise is deserved. If you are both making good progress you can introduce the whistle to get its attention, followed by the upraised hand and plenty of subsequent praise. Develop this response over greater distances and in different places. Remember you are hoping that, eventually, your dog will respond promptly to your 'Sit' whistle in the face of great temptation.

Sequence of 'Sit' exercises

With a young puppy use the command 'Sit', followed by good physical and verbal praise whenever it chooses to sit of its own accord. Incorporate the emphasised command in the praise.

1. **Introduce the command with gentle, slow pressure on its backside so that you can reward it as it actually *begins* to sit. If it sits, but gets up as a result you have another chance to repeat the exercise.**
2. **Require it to sit on command before feeding. If it sits, but gets up again as the dish is lowered, just lift the dish up again and make it wait. Eventually this can be developed so that the dog will stay sitting even after the dish has been put down – waiting for you to give the release command (probably the words you intend using for flushing, e.g. 'Get in').**
3. **Press-gang friends or family into 'running interference' by tempting the pup to disobey you – but never so energetically that it chooses them over you.**
4. **Introduce hand and whistle signals.**
5. **Always ensure you end on a successful note.**

Staying

The aim of this is that your dog will stay, in whichever position it is left in, until you tell it otherwise. It should not move because of boredom or distractions.

Sit stay

You have already been encouraging the puppy to think of sitting as being more than just an action. We want to develop this so that it becomes a position. By introducing the game of sitting at a distance, to the whistle, hand signal or command, the pup has got used to waiting for the next exciting bit. Being away from you is not seen as a punishment, and this is most important.

This pup is used to being gently unbalanced and is sitting very squarely in anticipation of this pleasant and reassuring contact.

Asking it to stay while you leave it, is the hardest thing to teach any dog. In contrast to the recall, the stay, in whatever position, requires a lot of confidence on the dog's part. You are asking it to stay away from its leader and divorce itself from the pack, both of which are alien to its mentality. Teaching a puppy to do this is even harder, as it is more dependent than an older dog would be.

Start by getting it to respond to a very soft command 'Sit', which we have already covered. I like using the sibilant hiss of the word, because it can be used when out hunting and provides emphasis without volume. A quiet 'Sssss' is enough to get a response from the dog, and is not obviously a human noise in the countryside when stealth is required.

With the pup sitting beside me, I give it lots of quiet fuss. Whenever it tries to jump up and lick me, I repeat 'Sit'. Simultaneously I stroke it, pushing it gently off balance, from side to side and back and forth.

No animal likes being unbalanced, and it is natural for the puppy to resist by setting itself more squarely. Ask a friend to push you gently off-balance in just such a way, and you will find that you have to instinctively resist. You can also use your foot or knee to stroke it, and it should resist this stimulus as readily. It will soon get the hang of this game, as you are doing this quietly, in an unexciting way, and the contact is pleasurable. You can even apply gentle pressure on the lead so that it has to resist being physically pulled. Your continuing praise and reassurance make this pleasant and rewarding.

This **contact conditioning** is a tool I use a lot in my training.

A first 'stay' lesson showing the sort of range of movement, from left to right, that a pup might be able to resist. Gradually increase your distance and the amount of sideways movement.

As it becomes more solid in the sitting position, I begin to move perhaps a metre away but returning very quickly should it think of moving from its position. Distance is unimportant at this stage. One of the best exercises is to walk around the pup. It will twist round to watch you, or to follow your movements, and will be tempted to move. At first you may only be able to move a little from side to side without it shifting. To keep it confident and happy you should maintain a continual commentary of praise while it is sitting. With practice it will get used to ignoring your movement and, eventually, will wait for you to resume your proper place by its side. (This exercise is the first I introduce when training an older dog.)

Develop this exercise very gradually by increasing the distance, and by including obstacles such as chairs between you and the pup. Even the momentary obstruction this provides will unsettle it a little. For a split second it could not see you properly.

Eventually you want to develop its obedience so that you can walk through an open door, and pause for a second or two before reappearing. As it gets better the time you can vanish for may be gradually extended. By now it will have a good idea of what is expected, and you can reasonably correct it if it moves. It will become more confident that staying is a good thing to do, and you can abandon the continuous verbal reassurance – but give more exciting praise to tempt it to move. Sometimes get friends or family members to talk to it. Or even to call its name very quietly – so that it becomes reliable in the face of this distraction. This is very useful where members of the public insist on trying to talk to an otherwise well-behaved dog.

With practice it will get used to ignoring your movement and, eventually, you will build up its confidence until, without any further command, you can walk away again – so that it learns that your return does not necessarily signal the end of the exercise.

A recent practical application will illustrate this. Some friends of mine and I had been training our dogs, and had adjourned to a local pub. It was my turn to buy a round so I left my dog in a 'stay', returning to the table with some of the drinks, before going back to the bar for the remainder. He stayed where he had been put while I walked back and forth. It did not occur to any of us to comment on what we thought was totally expected behaviour.

As the puppy gets used to you moving while it sits, you can *very* gradually extend the distance away from it and from side to side. Give plenty of vocal reassurance, which will keep its attention on you and stop it worrying about being separated from you.

You will often find that a dog that stays for relatively long periods of time will be tempted to get up and greet you on your return. This is because your absence seems like a punishment, and it associates your return with praise. By praising it while it is staying you will teach it that staying is a very positive exercise, and this will avoid the first misunderstanding.

To overcome the second it is a simple matter to stop short of it, and praise it again for staying – and then back off a pace or two. Equally we can return and make a fuss of the dog – but insist that it remains staying during the praise. Without any further command, walk away again, so it learns that your return does not necessarily signal the end of the exercise.

Throughout this description you will have noticed that I have not used the commands 'Stay' or 'Wait'. My belief is that a dog that is sitting cannot, by definition, be moving. The same applies with teaching the flat stay. You only add the second command because you doubt the dog's reliability to obey the first one. Nevertheless, I confess that, in common with every other dog handler, I do sometimes use these words without thinking – and subsequently curse myself for my lack of faith!

Generally, it is believed that it is uncomfortable for dogs to sit for prolonged periods of time and that if this is necessary, you should train it to lie down. A dog will certainly tend to stay better in the prone position, but my dogs resolutely stay sitting, when we are ferreting, for twenty minutes or so. Their concentration is absolute and they show no discomfort.

The flat stay

Use this position when you expect to leave the dog for any length of time (five minutes+). I use the command 'Flat', rather than 'Down', as it saves any confusion. It is an easy mistake to tell a dog to 'Get down' if it jumps up at visitors or gets on the sofa, yet fail to insist that it actually drops to the ground. Develop the duration and distance of this exercise as you would for the sit stay.

The stand stay

The stand stay is not normally used, except in advanced obedience competitions or the show ring, but is a very real exercise with a pointing dog. Teach it in exactly the same way as the other stays, with lots of contact conditioning (i.e. pushing gently from side to side, and backwards and forwards, so that it sets itself firmly) as you repeat the word 'Stand'. It will be extremely useful when you want to work on your dog's point. Practise it with the pup on the lead, beside a rabbit hutch or next to a pheasant rearing pen. Once it is established as a good and rewarding thing to do, it will make it very easy for you to approach your dog in the field. Because it does not associate staying with punishment, it will not feel pressurised to move and disturb the game. We will cover this in more detail when dealing with pointing.

Your dog should never be allowed to move from a stay position without your permission.

CONTACT CONDITIONING

This is a very useful technique in many exercises. It is based on exploiting the dog's natural desire to be balanced. The individual which is not balanced will be a little insecure, as it will be unable to protect itself or escape quickly. Pushing a dog off-balance does not scare it if this is done as part of a game. Its instinctive reaction will be to push back and thus keep itself balanced. Conversely, if we pull a dog towards us its reaction will be to pull away. By playing pushing games with our puppy we get it used to feeling able to push back against us. It can be done quite vigorously in a rough-and-tumble situation, or gently – when it creates a very affectionate response.

Once a dog is happy to push back against you, the technique will make it happy to be stationary. Pushing it gently right and left, forward and backward, while it sits beside you, will teach it to hold the sitting, lying or standing position – and enjoy resisting your pressure! The dog that sits a little away from you when recalling or retrieving, will move closer and counteract any attempt by you to push it gently away.

My own dogs will sit so close to me that they end up sitting on my feet. This is very useful as it means that I know exactly what they are doing – and where – while they know they are free to look around but cannot be left behind!

This resistance can also be used to make a dog stand back further from pointed quarry, and to be relaxed about your approach while it is on point. When it is pointing, it will be leaning as far forward as possible. Just push it gently further towards the game and it will naturally resist – but without feeling it has done anything wrong. In anticipation of being pushed beyond this point of balance, it will instead set itself more firmly to resist this.

Pulling him in creates resistance, and he pulls away. Push him away and he pushes back!

In the end he will choose to sit on my feet, so that neither of us can go without the other knowing.

At other times you may even use it to dissuade your dog from pulling on the lead – by pushing it forward so that it is having to slow down, or risk losing its balance by going faster than it intends.

Heelwork

Your belief should be that on the given command, the dog will stay at your heel irrespective of any changes in direction or walking speed and, when you stop, should automatically sit beside you. All this is implied by the one command – you should not need to add 'Steady', 'Turn' or 'Sit'.

Enthusiastic – but you would not want it pulling like this when it is bigger and stronger.

The well-behaved dog does not pull on the lead. It does not need to be yanked about when its handler changes direction, but should automatically follow him or her. Equally any change in your walking speed should be possible without the dog getting out of control or losing interest. This expectation applies just as much when the dog is not on the lead but has been told to 'Heel' or 'Close'. I choose the latter word because of the hard consonant it begins with, and to avoid any confusion 'Heel' might have with 'Here'.

Which side to walk?

The puppy should be walked on the same side at all times. Shooters habitually (but not if you are a left-handed shot) carry the gun in their right hand. It therefore makes sense to have the dog walking on the left-hand side.

Most falconers carry their birds on the left hand, so it would seem appropriate to have their dogs on the right. My own experience is that I always have a stick in my right hand, so the dogs prefer to walk on my left and suffer the indignity of the hawk messing on them. Make your own choice but be consistent.

We have already described how to introduce the puppy to wearing the lead, and to being taken to pleasant places on it. This means that it will have very positive associations with the lead and will walk forward happily into its collar. It does not take many days before it will start to pull, in anticipation of a free run or playing with other dogs.

Do not commence heelwork training in this situation, especially involving any form of negative control, as your puppy is being faced with the greatest temptation. Instead train it as an artificial exercise in the garden where there are no distractions. Only once it walks with its attention on you – because it does not associate this with going anywhere – can you venture out on the street. This must be achieved by making being with you seem the best thing out! Only introduce control when your puppy really does not do what you want.

Start when the puppy has become calm – why look for difficulties. Get it sitting by your side, and stroke and praise it until it looks at you. Once you have established eye contact say its name and then give a clear command. **Only then can you start walking.**

If you give the command at the same time as you set off, it will quickly learn to respond to your movement. This will result in a hiccup when you start training it to stay, as it believes your movement is the signal for it to follow! As your puppy will hope to be going somewhere exciting, it will leap ahead in its enthusiasm. You have two choices of technique here. Do you want to stop it pulling, or do you expect it to want to be with you?

Always start by getting its attention

There are two basic approaches. You can either give it a sharp jerk to bring it back beside you, at the same time giving a firm 'Heel' or 'Close', or you can simply change direction and praise the puppy for coming with you. I hope that the second is seen as the more constructive one at this stage. Repetition of this behaviour can become quite a game as you continually try to avoid your puppy's efforts to stay with you.

If you do this you will rarely get very far, as the puppy will be all over you. Calm it down at intervals by getting it to sit. Given good-enough praise it will begin to anticipate this as part of the game too, so that it will begin to sit whenever you stop.

Because of the excitement it will be jumping up and probably going too fast a lot of the time. Keep it fun and you should be able to turn sharply left or right, or to change direction completely, without needing to give the dog any command. A complete 180-degree turn in an anticlockwise direction will really test your pup's concentration. You should eventually aim to keep turning on the spot like this while your dog walks backwards to stay at heel. But always make sure the puppy believes this is a game designed for its benefit!

There will be times when the older puppy, or an adult dog, disobeys the 'Heel' command and pulls. Possibly you have found that the effort of keeping its attention on you has become too difficult to maintain. Equally it may want to go up to another dog, to reach the park more quickly, or a host of other expected pleasures. I beg your pardon! Who is the pack leader here?

Now is the time to use control. Your dog is questioning your authority and taking over. Correct it very sharply and firmly with a jerk on the lead and a fierce tone of voice. Leave it in no doubt that you are displeased. If you have been cross enough, it should be eager to placate you, and ignore the previous distraction. **As soon as it responds, change back to the pleasant, entertaining you. Aim always to make obeying you pleasant to the dog, instead of just being a way of avoiding correction.** Creating this contrast enables it to choose the response that is most acceptable to you.

Perhaps your control has not been effective, in which case I will need to describe how to jerk a dog. This will rarely be needed with a growing puppy but, with really headstrong youngsters or older dogs that have become wayward, you may be forced to employ tougher measures.

All training should be done on a slip lead for sensitive or young dogs (usually with a 'stop' system of some sort to prevent it becoming too tight), and a check chain for adult 'thugs'. With puppies the slip lead should be sufficient. Do not have any other collar underneath this equipment as it will interfere with the slip lead working properly.

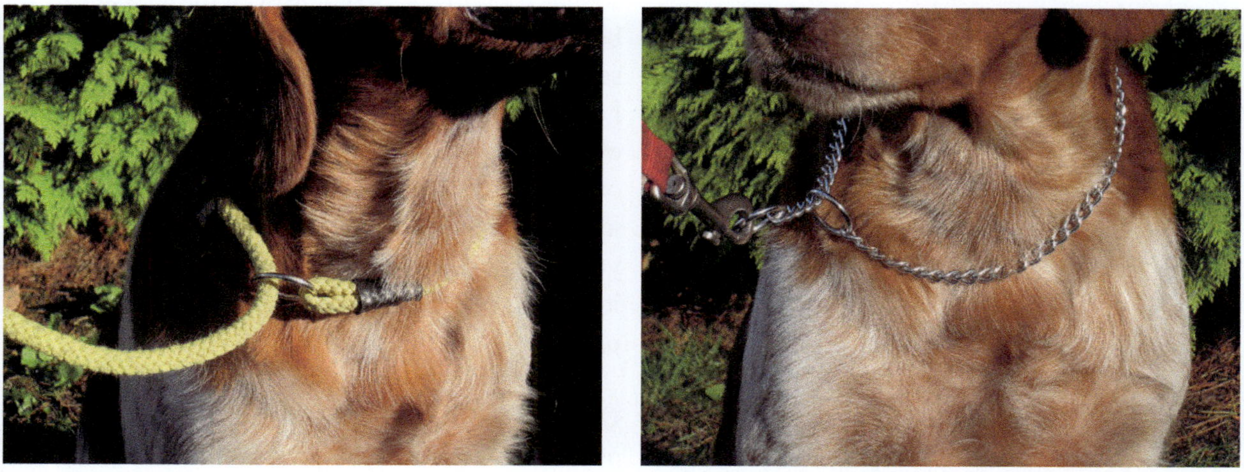

The correct way to put the slip lead or chain collar on. With the dog sitting on your left side, the lead or collar should flow in a continuous line from your right hand, across the top of the dog's neck, and finish under its throat. The weight of the large chain ring at the end pulls the collar or lead loose so that the dog is not being continually choked. For dogs worked on the right the process is reversed.

Jerking the dog

If you decide you will have to give a good jerk with the lead, learn to do it well, before actually putting your technique into practice. Most people have difficulty doing this, so it makes sense to master the art before trying it on your dog.

Place the noose of the slip lead or check chain, over a fence post and tighten the lead. Then pull it as hard as you can. Basically, nothing will happen except that you will tire yourself out. This is rather what occurs with most pulling dogs. Now let the lead go slack before snapping it tight as fast and as hard as you can. Quite possibly you will jerk the lead out of your own hand. This will teach you that jerking on a loose lead is much more effective than after the lead has become tight. Once the dog has started pulling you have lost this ability to correct it firmly. The best time to jerk the dog is before the lead has tightened.

However, dogs can move remarkably quickly, and you may have lost this opportunity. Do not despair. Drag the dog back towards you, and then release the tension on the lead, followed instantly with your hardest jerk on the newly slack lead. Since the dog will almost certainly be moving forward, it will add its own weight and momentum to your power, and create an even bigger jerk. Lady handlers, in particular, may lack the sheer physical strength to adequately correct a heavy dog with one jerk **(but did she pick the wrong breed and puppy for her physical strength in the first place?)**. No matter, a series of quick, determined jerks in very rapid succession will be seen accumulatively as a major punishment. The most important point to remember is that you have to win, no matter how long it takes. It may be a quick victory, or a triumph in a war of attrition, but you must be determined to be the boss.

This correction should be accompanied by a very firm command 'Heel' or 'Close'. You should stop dead in your tracks, or even step back a pace or two, so that your dog has to come even further back to get beside you. If the dog fails to come back to you, or sets off again immediately, repeat the exercise. It will not do this indefinitely, particularly if you reward it every time it does, fleetingly, end up beside you. Only once it stays beside you **until you decide to set off again**, can you claim that you have been effective. If your dog is cowed by this you will be able to praise it for not pulling, and thus turn the punishment into a pleasant and positive choice.

One of the commonest failings with heelwork occurs when the handler corrects the dog for pulling and gives firm commands, but continues at the same pace without making the dog check in its stride. How is the dog to understand what you mean when nothing has been changed?

Five weeks into her training – and no longer pulling like a train but relating to her handler – and the regular free-running field is only about 50 metres ahead.

Be effective. When you stop, it has to come back beside you, and you will not set off again until it shows that it is settled in this position. It can be a battle of wills but you can and must win it, and since you are holding the lead you will do.

Many dogs display a form of heelwork evasion, whereby they walk ahead of their handler's knee on a slack lead. This enables them to watch out of the corner of their eye, so that they can still turn with him or her. Doing lots of sharp left turns and left about-turns will force the dog to keep a pace further back. Also alter your pace so that the dog has to watch what you are doing. Push the dog away in play as it walks and it will push back into you, or even jump up as it walks. It will be truly engaging with you if it does this.

HPRs may be using their nose while they are at heel (in fact almost certainly they will be doing just this). Ideally you want eye contact but their minds are on other things without actually being disobedient. I find a light tap on the back of their head is unexpected and makes them look round. If they had been looking at me, I would not have been able to sneak up like this, so they have to look at me more. Then, lo and behold, they get lots of praise. Practise your heelwork before any training session and include, as you get more confident, doing this off the lead. Given one command, a well-trained dog will keep to heel irrespective of any turns or changes of pace, and should sit automatically when you stop.

This old photo shows my first two Brittanys walking at heel, off the lead despite the obvious temptations. The black and white bitch was my foundation bitch, from which nine further generations have come.

If you have friends or family available, get them to start introducing distractions during this and other exercises, so that the dog learns to ignore other things and is truly obedient. Weaving in and out of other dogs is a good test. Like all training, the object of the exercise is to get control of the dog's mind, not its neck. You will know you have this when your dog looks at you confidently and attentively, waiting for your command **and your praise**.

Recall

This should be the easiest of lessons to teach as the dog should love being with you. Sadly, it is often the hardest because we wait until it has become independent, and possibly a bit disobedient, before we start. By this time we are playing catch-up, and trying to remedy a rapidly deteriorating situation.

The recall falls into three parts:

1. **Teaching it that being with you is fun**
2. **Teaching it not to run away**
3. **Teaching it to come when it is called**

I should not need to insult your intelligence by suggesting that the dog that enjoys being with you and does not run away, is not a problem to get back.

Staying with you

The recall is in itself a negative exercise. It assumes the dog has gone, or will go, further from you than you want. Much of good training consists of preventing bad habits, so it makes sense to focus on making the dog want to stay close to you from the start.

Dogs are by nature gregarious. They like belonging to a pack. In the domestic dog the pack is your family. The puppy wants to be with you most of the time, and your job is to make this a pleasant and rewarding experience. Talk to it and fondle it. If it curls up beside you just rub it gently with your foot in acknowledgement of its welcome presence. Make eye contact with it, and get it used to responding to your varying tone of voice. At any time, when it trots up or bounds to meet you, do give it a quick pat or a kind word. It is easy to dismiss it as 'that pesky pup', when all it wants is a bit of affection. When you take it out for its exercise, concentrate on getting its attention with your voice and your whistle, and backing this up with lots of praise when it comes to you.

Incorporate the dog's name and the command 'Come' in its praise. It may get to the point where the pup seems to think that the whole purpose of going out is to have a rough and tumble with the 'boss'. Do not worry, it will soon start exploring wider areas, particularly when it gets switched on to the scent of game.

Once your pup is doing little retrieves you can start using this too as a reward for returning to you. **Recall is not the end of freedom, but an invitation to come and do something with you!**

Not running away

As your puppy gets bolder and more adventurous, you can introduce the game of hiding from it. Wait until it becomes fixated on a fascinating smell, then quietly sneak away in the opposite direction. It will quickly miss you, and will race after you at top speed – to be greeted with a suitably rapturous reception. Extend this exercise by hiding from it behind trees (always dubious behaviour in the eyes of the local neighbourhood watch), or dropping flat in long grass.

Not only will it panic a little (which is good), but it will also start learning to use its nose to find you. As the pup gets more proficient at discovering your hiding place, you can get more ambitious. Climbing trees can work very well, as young dogs do not think of looking up. From your vantage point you can watch as it becomes more and more worried. Its ears and tail will go down. It will run in one direction, then another. It may even start whining in its distress. Only whistle or call if it looks as though it might go too far. Your signal will draw it back towards where you are and stop it getting into danger. Then keep quiet again. Eventually it will find you.

What it has learned is that it must keep an eye on you, or risk getting separated from the pack. It also has learned that if it perseveres with scenting, it will eventually find you. This will teach it to keep hunting out blind retrieves and wounded birds later in life. By the time hiding no longer undermines its confidence, because it has got so skilled at finding you, it will have got into the habit of staying in touch with you and responding to your call.

Although this forest is vast the puppy wants only to stay with me. He is confident enough to explore light cover and to investigate scents, but I can easily sneak off and hide from him behind trees. He will learn to be constantly aware of where I am while he is working. Praise whenever he does find me will continue throughout his life, and will help maintain a really positive recall.

Coming on command

There will be times when you want the dog to come away from an interesting smell, unauthorised ground or potential danger. You need it to obey instantly, and this is where a consciously taught response is essential. This is one exercise which must be non-negotiable and, ideally, it is taught from very early days. I mentioned earlier that the first thing we teach our pups are its toilet habits. Being taught at such a tender age these habits are extraordinarily hard to change. They are virtually imprinted on the dog's mind. It makes sense to start recall training at an equally early stage.

My own technique is vastly different from anything I have ever read in any other dog-training manual. It was developed as a result of bitter experience with my first dog. As long as I can remember my family bred dogs, in fact I think it was a lasting regret of my mother that I had only two feet and no tail! She had a favourite bitch from which she produced a litter. Dad decided he would keep a pup, and chose a very bold bitch. Within a few months he decided she was too difficult to control, and gave her to my sister. She, in turn, quickly became disillusioned and passed her on to me. I was the youngest so was stuck with what was, by then, a fairly hardened criminal.

Over the following years we achieved limited success at obedience, but she became increasingly wayward when free running. She was a committed escapologist, and what the Americans call a 'running fool', often going missing for up to forty-eight hours, endlessly hunting for game. By the time I left home to start work as a professional dog trainer, she was thirteen years old. Even at that advanced age my mother had to have her put down, as being beyond reasonable control. I swore I would never again have a dog that was bad at recall, and began to evolve a reliable method of teaching this vital exercise.

The most frustrating dilemma for any dog owner is that they cannot punish the misbehaving dog when it finally condescends to return. Any correction will only be associated with coming back to you and it will, therefore, be less inclined to return next time. Unfortunately, it is too far away for you to be able to inflict any punishment on it while it is ignoring you, so it is in a no-lose situation. It can enjoy its freedom and still get praised (probably through gritted teeth) when it finally decides it is hungry or tired enough to come back. We need to find a way to punish it while it is out of range. There are devices available that will, from a distance, give an electric shock to a dog, but these are socially questionable and should have no place in training a puppy. Also, some dogs learn to be obedient only when they are wearing them (or substitute dummy collars). My technique works on a psychological level.

Disciplining yourself

Before outlining the procedure I use with puppies, and also describing how to extend this to older dogs that are already steeped in sin, I should make something very clear. At the outset you are going to be absolutely horrid to your dog. You will undermine its self-confidence and thoroughly upset it. If you are not prepared to do this effectively you will not succeed. However, you and your dog will quickly realise the benefit and will enjoy a much happier life together as a result.

The other warning that must be given is that while you will give every appearance of being extremely angry with your dog, you must **never** do this if you are genuinely in a temper. The success of the method depends on your ability to switch from apparent fury to ecstatic joy in an instant. The clearer you make this contrast the more readily the dog will understand what you want of it.

The first signs of trouble start when the puppy is still in the garden. There will probably be an occasion when you want it to come in, and it decides to scamper about or investigate more interesting smells. At this stage it is easy just to stride over and pick it up. Unless you are very careful this will develop into a game, with the puppy getting more and more expert at avoiding your outstretched hands. What fun. This is the time to teach your puppy its first recall lesson – but only if you have ten minutes or so to spare. Otherwise wait until you have nothing more pressing to do.

Every dog has its 'flight area'. Within this it has to decide whether new things are potential dangers, and if it should run or fight. All dog owners recognise this in their pets. An understanding of this will help you teach the recall. When your puppy first refuses to come, you should casually approach until you are at the edge of its flight area. At this point it does not have to do anything, but it is aware of your presence. Suddenly, in a loud, deep, angry voice, say 'Come!' With an invitation like this your puppy will immediately bunk off a bit further, hiding its confusion by pretending to play, or by inspecting a particularly interesting flower, etc. It has recreated its flight area and can ignore you again. Once more casually saunter up to the edge of the area and repeat the command as fiercely as before. Expect several repetitions of the puppy evading you.

What you are trying to do is, simultaneously, to undermine its confidence and stop its enjoyment. It does not fancy coming to you and yet it cannot enjoy itself. Equally you are making no effort to catch it, so it is in a situation that it does not know how to resolve. As its self-confidence evaporates it will increasingly want reassurance, but will be frightened to come to you. Let it stew for a bit until it is looking pretty pathetic and upset. It will not come to you, so when you judge it is being very submissive, you will have to grasp it gently under its chin (it will probably be lying down) and drag it firmly towards you, repeating 'Come' fiercely and backing away as you do so. This is when you suddenly reverse your behaviour and praise it ecstatically, including the obeyed command in your praise – 'Good puppy, **come** – good puppy, clever puppy' and so forth. Once it has perked up you can let it go again so that the lesson can be repeated. Do this perhaps three times and it will start coming very quickly and confidently.

It will only take a couple of these little sessions to teach it that you can stop it enjoying itself at a distance and, most importantly, that it can get pleasure and relief from this predicament by coming to you quickly. It will have found that the 'Come' command is non-negotiable and an imperative.

The advantage of training the recall like this is that the dog knows you can spoil its fun, even when it is out of reach. Also, because it is accustomed to a firm command you will be able to shout at it, and get a prompt response, in moments of crisis. I defy anyone to use a happy, cheerful command (as recommended in most dog-training manuals) as their dog races towards a main road, or sets off into a field of sheep.

Older dogs can be trained in exactly the same way, but it may take longer and it is sensible to start in a fairly controlled environment such as a smallish sheep paddock. People will think you are nuts as you wander after your dog, bellowing at it in a furious voice, but making no attempt to catch it. Let them. Your job is to spoil your dog's fun and create an anxious state of mind.

As an apprentice guide-dog trainer, I spent an afternoon doing just this, under the uncomprehending gaze of my boss, who watched from his office window. He was waiting to tell me off for losing my temper with the dog I was apparently furious with. It took me two and a half hours to finally get it to give in and come to me. I praised it like mad and let it go again. This second time it took thirty minutes and I again let it go, after a play session and some heartfelt praise. The third time only took five minutes and it refused to leave my side after that. For the rest of its life it was excellent at recall but refused to enjoy any freedom at all, rarely going more than a few metres from its owner's side – because it had learned not to go away!

If you are still confused by the psychology of this, I suggest you go to your local dog-training club. There you will see dogs being told very firmly to stay. Equally you will see several of them start creeping towards their owners in an anxious attempt to placate them.

General obedience

General obedience is something that is required of the dog throughout its life and in every work, play and social situation. Teaching it skills or tricks does not make it obedient, but it is a good start. Obedience gives you the power to make the dog behave when it does not want to. It should leave fascinating smells or discarded food containers. It should sit when told, even if game is being flushed in every direction. It should come back on command no matter what temptations may exist, and it should do as it is told even when not working, to 'give over' as my father-in-law put it. The dog should never question your authority.

If you are one of the many dog owners who say 'He'll do it at home,' then you have a trained dog but not an obedient one. If your dog is thoroughly well behaved and biddable, but does not know how to 'Give a paw', fetch a wounded pheasant or do a blind retrieve over water, you have an obedient dog but not a trained one.

Look for small instances of disobedience or lack of respect for you. Does it go through doors in front of you? Does it refuse to sit for its food and, instead, nearly knocks the bowl out of your hand in its eagerness? Does it come back when it fancies? Does it pull on the lead to get to the park or to see another dog? Does it beg at the table? Is it noisy when you leave it?

Most dog owners choose to overlook these little transgressions. Don't be lazy – it is worth picking these points up from time to time just to remind the dog who is boss – and to reassure yourself that you still are!

Evasions

Dogs are very clever at managing to disobey in little ways. We call these evasions. Just yesterday I watched someone practising the recall with an unruly young dog. It would come promptly, but repeatedly danced off to one side or the other rather than stopping beside its owner. It evidently felt that it had come as instructed, but saw no reason to stop its play. The owner also believed the dog had obeyed him, yet was aware he had not got the response he wanted. If your dog or puppy does this, just back off further in the opposite direction to that in which it has jinked away. Keep doing this until your dog gives in and comes to you directly.

Another evasion occurs when a dog delays returning by stopping to investigate an interesting smell. It is not that it has refused to obey, just that it has not been as prompt as it should. Parents will find it easy to identify with this. Training the recall as described above will remove the dog's apparent belief that this is optional.

In a way it should not be difficult to overcome evasions, as the dog has demonstrated that it lacks the courage to actually disobey. All you need is a clear idea of what you think is implied by each command, and a determination to achieve this. Evasions are symptoms of insufficient respect, so be on your guard for them.

Permitted evasions

There is one constant exercise in the pointing dog's repertoire which should be seen as a permitted evasion. When it is hunting, you use a single *'peep'* on the whistle to turn it back towards you, but you do not actually want it to come completely back. Instead we give a weak signal and the dog turns back across our path in a quartering pattern as it continues hunting. We reward this with praise and the dog also gets rewarded by being allowed to indulge its chief urge – to continue hunting. When we want it to return, we have to give a louder, longer signal, *'Peep – peep – peeeep'*, to emphasise that, this time, we really mean it.

Experienced working dogs develop a tremendous amount of initiative and will often anticipate situations in an attempt to find or retrieve game. An example may be where a dog extends its outrun when quartering to take in a patch of cover which we had not directed it to. We do not want to discourage such enthusiasm. If we are sensible, we respond positively to these instances, as they are attempting to please us and are not being wilful. But they are questioning our authority and we need, occasionally, to pull them up and make them do things in the way we want. Otherwise they will start to take more and more liberties with our lack of control.

Chapter 5
The Retrieve

Caveat – this chapter covers several months of your pup's education, and you should not try to cram it all in at the start. How quickly your pup progresses will depend on many factors, and is not some sort of a race. You will never stop teaching this exercise because real-life experiences (and imaginative judges) will keep coming up with unexpected variations to be practised. A very common training mistake is to keep challenging the dog's understanding by making the retrieves increasingly difficult – until it fails! Lessons learned by success are much better than those ending in failure.

If you are reading this because you have a pet HPR, do not dismiss retrieving as something you do not need to teach it. A dog that enjoys retrieving will be easy to exercise vigorously in quite restricted areas. Its enthusiasm will mean you can reward it with a retrieve whenever it returns to you. Being with you therefore becomes more exciting, because you can entertain it more fully. Develop this skill until it can find its favourite toy in bushes or long grass. This will exercise its nose and brains as well as its legs! You will be engaging with your pet, and may view it with more respect when you discover how talented it is.

Of the three definitive functions of the HPR group it is fair to say the retrieve is the one that is most likely to need specific training. Some dogs take to it very naturally, but few HPRs have the same compulsion to carry things that you would expect from breeds such as Labrador Retrievers. All puppies have a need to play and will chew at all sorts of things. Within the litter they are normally in competition with their siblings, so it is natural for them to be both possessive and destructive with their toys. Neither of these qualities is really what you want in your working dog. What you dream about is having a dog that slavishly brings everything to hand with the softest of mouths.

In the first place, accept that if you are a shooting person, it is better to have a dog that brings things back than one that does not. Secondly, it is better that your dog brings things back that at least bear a passing resemblance to the creature you brought down. Of course, if you are a falconer the retrieve is immaterial as, when the hawk catches its quarry, the dog has no further part to play in the drama. In fact, you will encourage your dog not to retrieve, as this might lead to it chasing flushed game. Some dogs are expected to work in both disciplines and seem to know perfectly well when the retrieve is relevant and when it is not.

Your puppy will not share your lofty ambitions. It is more intent on shredding the doormat, reducing lumps of coal to coal dust or splintering wood into the tiniest matchsticks. Where do you start?

Play

Your puppy does not share your hopes. Its life centres on play and food and, quite properly, it needs to check the edible qualities of its toys and the play potential of its food. To some extent you can prepare for the future by making sure that its toys are fairly indestructible so that it never gets the satisfaction of ripping things apart. Such behaviour, carried over to its working life, would not augur well for the condition of any game it retrieves.

Choice of toy is worth thinking about. It must be soft enough to be attractive to the puppy but tough enough to resist needle-sharp infant teeth. The weight should be easy for the pup to carry. Avoid anything that might be hazardous such as squeaky toys, plastic or metal objects or anything that might break into smaller, sharper pieces. **Squeaky toys are particularly unsuitable** – as they encourage a hard mouth in puppies, which will want to make them squeal.

Some retrieve articles are positively dangerous. Balls, other than fabric-covered tennis balls, should not be used. As the pup grows it may get a ball, slippery with saliva, lodged in its gullet. Your attempts to remove it will simply push it further down and, unless you carry out a canine-style Heimlich manoeuvre, may choke it to death. Stones can be swallowed, leading to expensive vet bills in the short term and worn teeth in later life. Sticks which may splinter, or which have jagged ends, are also potentially lethal.

During play the puppy will originally have been in competition with its litter mates. Naturally any toys become subject to disputes, and it is normal to see pups tugging at opposite ends of an old jumper or bit of sacking. Interesting pieces of wood (see above) or bones also get purloined from one to the other and the victor will race about, shielding its prize from the competition, until it can find a quiet corner in which to exercise its young teeth. **Jealousy and possessiveness** are natural but not wanted in the future. So do not let the puppy introduce games where you have to run after it to get things back, or have to pull its toys away from it.

As much as possible you should try to create the situation where you can sit down beside the pup when it has a toy. Make much of its possession, and admire it, but do not try to take it away. The pup will learn that having something is praiseworthy, but that you will not try to steal it for yourself. You want to get to the stage where you can sit down and the pup will come over to you, bringing whatever it is chewing with it.

Quite possibly your praise will distract it from its toy. Keep fussing over the pup but throw the toy a metre or two away. Almost certainly your future star will bounce over and grab it. Use your voice to make being with you attractive, and it may well return with the toy in its jaws. **Do not try to get the toy from it**. Just concentrate on the pup feeling that being with you is fun. In anticipation of your affection it will be less interested in the toy, and you may be able to purloin it again for another little throw. Because you can make the toy do interesting things, you become an integral part of its play, and your puppy will want to exploit you by repeatedly bringing the toy to you to be thrown.

Some puppies can be very inhibited about picking things up, as a result of being chastised for having stolen, or chewed, forbidden items. This clearly creates a discipline for you. Make sure that it cannot get hold of valued items by keeping only permitted toys within its reach. Even pups that have developed this concern can be got to play with toys, if you are prepared to 'mock-play' with these yourself. If you show particular excitement when you roll a toy around, the pup will reason that the toy is incredibly interesting. The competitive nature of puppies will encourage yours to get involved.

Develop this play so that the thrill of bringing things back to be thrown again, becomes more ingrained. Introduce a variety of different objects, and gradually increase the weight. Make the retrieve more challenging by throwing the toy behind the settee, or through an open door, so that the pup has to use its initiative (and possibly its nose) to find it. Even at this stage it will begin to listen and **'mark'** where something has fallen. Always ensure you give really good praise to keep stimulating its enthusiasm. As the game is practised more, it is the chasing that becomes more fun and is the most anticipated bit. Getting the pup to release the toy gets easier, but always be prepared to play until it is ready to give it to you, rather than prising the toy away prematurely.

Chasing and hunting

The games you have been playing in your sitting room will continue in the garden and when you start taking the puppy for walks. It will really enjoy the thrill of the chase but, of course, in the wild this would be merely the precursor to the kill. Accept that at this stage it does not matter that your pup is racing after the retrieve object without any form of control. There will be time enough for that later. Just exploit its keenness by very gradually making the retrieves longer and more difficult. Move from short grass to grass that hides the object. The pup will have no difficulty facing this longer grass and it will quickly learn to **'cast about'** to find the missing toy.

This half-size dummy is barely hidden but has clearly been scented by this thirteen-week-old puppy

You can help by joining in and encouraging it to look in the right areas. When later you want it to search deeper cover it will be used to responding to your urgings, and to expecting a positive outcome. If you fail to generate any enthusiasm or to become involved, this will be reflected in your dog's future working style.

As the pup grows in experience it will search more persistently and can be accustomed to increasingly rougher and deeper cover. My latest pup, at five months old, will go into thick bracken and heather, but I do not ask her to cope with dense, thorny thickets. She will fetch dummies thrown into small, thin bramble patches, but I am careful not to over-face her as she is quite sensitive.

Seventeen-weeks old and very happy to retrieve from moving water

Virtually all her retrieves now drop out of sight and she has learned to use her nose and to cast about pretty well. Sometimes she misjudges the distance and I have to encourage her to the area. However, although she is becoming familiar with them, she does not understand directional signals yet. She does persist for several minutes until she is successful at locating the object. She is also retrieving well from water, having started tackling shin-deep trickle at about three months and working up from there until she now tackles quite swift-flowing streams with confidence.

From an early age I begin using dead game for retrieves. I am fortunate that I keep frozen quail and half-grown rats to feed to my hawks. Both of these, when thawed, make an excellent starting point as they are small enough for a puppy to carry easily. Too heavy an article can put a pup off. It is important to get it used to the texture of feather and fur in its mouth, and these little carcasses also give some scent when they are out of sight. I put my pups on the lead and let them carry the bodies triumphantly for as long as they want. As soon as they drop these, I give lots of praise, and they will take them from my hand again. After a second or two I then reclaim them, safe in the knowledge that the puppy is ready to relinquish its prize.

Very obviously there is a progression towards the persistence with which a pup will eventually hunt, which has been built up by a pattern of success. Writing about it makes it sound easier than is sometimes the case, and you may have to retrace your steps to regain its confidence.

If your pup is reluctant to readily hold a whole carcass, no matter how small, try fastening a few pheasant feathers, progressing to whole wings, around their favourite toy or dummy. Do the same with a piece of rabbit fur. Nylon cable ties or insulating tape are ideal for this purpose. Once this early reluctance is overcome, the article becomes even more desirable. Be careful, however, that the pup does not begin to stop and chew the article. This is why you must have a good retrieve *before* introducing game to the 'menu'.

A dead quail is ideal for this sixteen-week-old puppy

You will soon be able to vary the retrieve articles and to introduce **cold, dead game**. This is a middle step before using warm, fresh game, which has such a powerful scent that pups can be quite nervous about actually picking it up. Add the unexpected texture of fur and feather and you may find your pup is really hesitant to touch it or get its teeth round it. Be heartened by this reluctance as it is associated both with a strong tendency to point, and with having a soft mouth.

Very obviously there is a progression towards the persistence with which a pup will eventually hunt, which has been built up by a pattern of success. Writing about it makes it sound easier than is sometimes the case, and you may have to retrace your steps to regain its confidence.

Control

Everything so far has kept in mind the need to maintain a high level of motivation. If you have succeeded, your pup will be a 'retrievaholic', and something of a pain in the backside, continually demanding that you throw something for it. This is important as, when you introduce control, the pup will be confused and a little put off. It will perceive any discipline as a punishment, **so make sure it is equally as happy to sit and stay as it is to retrieve.**

Begin with doing a short obedience session, i.e. some heelwork, and making it sit on command and stay. It may be anticipating the much more exciting retrieve, because it will have smelt that its favourite toy is in your pocket. Nevertheless, it will settle for this lesser amusement.

Once it is doing a sensible stay, put it on the lead. Then tell it to 'Sit' and throw its dummy or toy a short distance. If it tries to rush after it you can give a light correction before reverting to the usual praise for staying. Wait until its attention is back on you, and it has almost learned to ignore the retrieve article. Only then should you remove the lead, keeping it staying with your voice, before telling it to 'Fetch'.

Having done this once, return to the 'Sit' exercise and the next time, leave the pup sitting and go and fetch the article yourself. This will be highly confusing behaviour and a reversal of every known sequence of events. Keep alternating the two, with you doing the majority of the retrieving, so that the pup is unsure whose turn it is. The intention is that the pup will be prepared to wait for your command before racing off to fetch the article. The distance involved should be pretty short, as the pup needs to be close enough to more readily accept your control. Once it gets further away it is much more likely to 'stick two fingers up at you'!

Once your puppy is fairly steady and sits reliably while you throw the retrieve (possibly between five to nine months old depending on its progress), you should develop its steadiness further. Try to throw the dummy or article while the puppy is moving about close to you – even just tossing it up in the air before catching it yourself. As the article becomes airborne say 'Sit' very sharply, or give the whistle signal for this. You want the puppy to associate the movement through the air with sitting, not chasing (which all its previous experience has been). This is an easy introduction to the concept of '**sitting to flush**'.

This is a major reversal and can cause it some confusion. For this reason I like the youngster to be pretty keen before introducing this. Other trainers will recommend establishing the control first, and your own choice may depend on your personal assessment of your puppy's temperament. Basically, work on the control early with 'pushy' pups, and concentrate longer on the enthusiasm with sensitive ones.

Probably the greatest bane of the HPR trainer is the ever-present rabbit. You and your puppy may meet them at frequent if unpredictable times on your walks. They have a most unfortunate habit of running away, which your pride and joy will find irresistible. In the shooting field your dog may well be at risk from a trigger-happy gun with ambitions of having rabbit pie. Falconers will find it a real nuisance too, as a chasing dog may well put off the hawk, apart from speeding 'bunny' on his way a little too quickly.

Your control can be rapidly eroded by rabbits, and you do not want your young dog to get into the habit of chasing these. However, if it is not particularly obsessed with retrieving there is a risk that, in correcting it for chasing, you may also discourage it from running out for a retrieve. Your work in teaching it to sit, as dummies are thrown, is a good starting point. It will have learned that doing this is desirable, because of the excellent praise it has received.

Now set up an artificial exercise where you arrange for an assistant to pull a dummy covered in rabbit fur, or a dead rabbit for that matter, across your path while the dog is walking at heel. Whether the dog is on or off the lead will depend on your optimism and confidence. As soon as the dog spots the movement tell it sharply 'Sit', or give the stop whistle. This must be done instantly so that your dog has no time to get up any speed. You can now turn the exercise into a positive one, and can reward it for obeying your command. Eventually you want the dog to sit in anticipation whenever a decoy is pulled across its path.

The ultimate reinforcement of this is to take the dog ferreting with it sitting, tethered initially, beside the warren. Here you have the smell of live rabbits, the sound of their feet drumming underground and then the quick bolt from the hole. Again, a prompt command or whistle will encourage it to ignore the rabbit and stay for your praise.

Directional commands (refer to Chapter 8 for more information)

When your pup sits fairly reliably, you can leave it sitting and throw a dummy a few metres to one side. The pup will be watching that, rather than you, so **wait until you have regained its attention** before continuing. Then go two or three paces in front of it, so that you are facing each other. Use quiet praise to keep its eye contact on you, and to keep it on the same spot. When it is concentrating on you, give the command 'Fetch', with an exaggerated arm/hand signal in the appropriate direction. Repeat the exercise on the opposite side.

You will already have used these signals when free-running the pup, and it will pick them up quickly because it is highly motivated by the retrieve. As you gain confidence in your pup's obedience you can increase the distance you stand away from it, and how far the article is thrown.

The hardest signal to teach is the one that asks the pup to go further away from you. It is given by holding your hand up above your head, as you did when teaching the dog to sit at a distance. By gesturing towards the dog in a pushing motion, we try to communicate our wish that it goes away (see photo).

Go (to my) left signal and posture

Go (to my) right signal and posture

The 'Go back' signal

If you remember the psychology involved in teaching the recall, you will realise that being negative will only make the dog want to come to you. Providing a good-enough incentive is key to your success.

I start this exercise by surreptitiously dropping the dummy behind me as I walk along the side of a field. This placement is chosen because the fence or hedge line gives the pup a direction to follow. Ideally it should be done along a fenced footpath, where the pup can only go forwards or back. The first time I try this I only ask the pup to go back a few paces and find a dummy that is clearly visible. It does not take long to extend this distance and to achieve this with the dummy hidden in long grass or light cover.

The next progression is to get the pup to go ahead for an unseen retrieve instead of retracing its steps. Start with a few retrieves and stays before trying this. Then put the pup back in the car, before deliberately placing the retrieve article perhaps twenty paces ahead. Get the pup out and make it sit facing you with its back to the unseen article. The same signal is used, perhaps with the command 'Back'. It may be confused, and you will need to encourage it onwards until it stumbles across the article. Loads of praise is in order and you will be able to slowly develop this exercise. It is also a good idea to commandeer an assistant to throw a dummy well ahead of you, but within the dog's range of vision. Now make it wait, facing you, before giving the signal. It will quickly learn to race ahead to seek anything retrievable. Incidentally, if it brings back something else, simply reward it and send it again. Once it gets switched on to game it will soon learn to concentrate on that to the exclusion of everything else.

If you have got the pup to the point where it will go ahead, and then stop on command and wait for a fresh signal, you can really feel proud. You can now start sending it out into open land and get it to work over selected areas. Well done.

Split retrieves

As a way of testing your control, and your dog's ability, it is worth introducing split retrieves. This is where two or more articles are thrown in different directions, and you are able to send the dog for whichever you choose. This is a required skill if you have two or more birds shot, of which one is a runner. Obviously, this must be the first one to focus on, even if it was not the most recently brought down. Success at this requires excellent control from you, and great willingness from the dog. Habituating the dog to staying while you pepper the immediate area around it with dummies, is not really appropriate for HPRs. It is taught as a preparation for 'peg dogs' on driven shoots, which is not what they are about. You can, if you wish, do this, but not until the dog is quite experienced, possibly approaching a year or more.

Some dogs are tempted to drop their dummy or bird when spotting another lying nearby or, even more exciting, one which drops while the dog is already in the process of retrieving. This is a fault and you can easily simulate this by throwing a dummy out beside the dog's line of travel while it is returning. Once it learns to ignore the second dummy you can, as soon as it has delivered the first, send it for the second and, as it approaches this, throw the first out in turn. This can be done repeatedly, provided the dog is keen enough, and it will soon learn to ignore the extra dummies until it has delivered the one in its mouth. It also speeds the dog's delivery of birds – as it is in a hurry to go for the next!

This is a good exercise to practise as it reinforces your control while at the same time encouraging a prompt delivery to hand, as the dog is anxious to go for the next one. This can be increased further by making the dog do a few paces of heelwork with the bird or dummy in its mouth, before taking the article and then sitting it and redirecting it to the next, selected one. You can also use several dummies or birds to complicate things as it gets more skilled. It also underlines that the fall of a dummy or bird is not the signal to retrieve. It must wait to be directed by you. Because the reward element in the retrieve is so great, your dog will be extremely willing to achieve what you want.

Blind retrieves

Very often the working HPR will be asked to retrieve fallen birds that it has not seen drop – so has been unable to 'mark' the spot where it came down. You need to be able to send your dog in any direction – for sometimes considerable distances – and possibly to redirect it if it has gone slightly offline. This involves being able to stop it and move it to the left and right, as well as bringing it closer (if it has overshot its target) or being able to send it further back. To perform these manoeuvres it must understand and respond to the signals, whistles and commands for going, stopping, right, left, back and come. Wow! When sending for a distant retrieve you should always 'aim' the dog to a point downwind of the anticipated dropping point. Stopping it too early will prevent it having any chance of scenting the fallen bird.

These skills accumulate with practice throughout the puppy's training. They start with your puppy being keen on the retrieve, and being prepared to hunt for an article in cover of some sort. At this stage we are not looking to introduce directions, but just to get the puppy to go away from us in a straight line. There are two different techniques for this and you will use both, but they have slightly different advantages and drawbacks. Let me describe them both, so that you will understand me better.

Option 1

Start in an open field but about ten metres out from the hedge. With the puppy sitting beside you, get an assistant to throw a dummy to the foot of the hedge. Then, instead of just sending the puppy as usual, line it up and send it along the direction of your hand.

The 'Fetch' command and signal

Of course, it has no idea why you have done this, because its concentration is wholly on the dummy it has seen drop. It should race out as normal and bring the dummy back positively while, unseen by your returning pup, your assistant has dropped another dummy in the exact same spot.

Repeat the exercise immediately and the puppy will almost certainly anticipate another retrieve, and will investigate the last place where it found one. It has not yet learned that your pointing means that there will be a dummy specifically in that direction, but it will associate your behaviour as another indicator that it will be sent for a dummy.

A few days of repeating this exercise in the same place and you can try sending it for a hidden dummy without first having a seen retrieve. It associates everything you do with finding something there, and will almost certainly bring back the dummy. As it gets more confident you can slowly extend the distance so that the pup is thinking of going as far as it can, i.e. to the indicated hedge bottom, in the expectation of finding a dummy.

Option 2

This is started in a retrieve lane, or a long, narrow space such as a fenced footpath or between rows of fruit trees or raspberry canes. Start with a seen retrieve thrown by an assistant, and with lining the pup up in exactly the same way as Option 1. Similarly, a second, unseen dummy is then thrown out by your assistant while the first is being brought back, and it is reasonable to expect that the pup will return there and successfully find the second. With practice you can increase the distance of the retrieves and send it for dummies without having first done a seen retrieve.

The two techniques seem very similar and both work well. The difference is that the first teaches the dog to keep going to the furthest visible point by working back from it, while the second teaches the dog to go further away from you and to anticipate finding a dummy on the way. If your dog tends to lack drive it is worth concentrating on using Option 1 to develop the confidence to go as far as it can see. Alternatively, with dogs that have plenty of drive, but not much steadiness, it is worth using Option 2. They will set off anticipating the presence of a dummy from the outset, and will be concentrating on their nose more. You will use both techniques as the dog's confidence increases or wanes.

Redirecting for blind retrieves

Once your young dog is going out pretty confidently, and fairly straight, for blind retrieves you can slowly introduce redirection. Here I would definitely recommend Option 1 as your starting strategy.

Before training the dog in redirection it is important that you can stop it at a distance (no further than fifteen metres will be sufficient to begin with), and that it is used to following hand signals to left or right (refer to Chapter 8 for more information on this). Now place a couple of dummies some thirty metres apart at the foot of the hedge and either side of the place you normally send your pup to for its blind retrieves. With these hidden, send your pup along a central line to where it expects its dummy to be. Not finding it in the anticipated place it will probably be undecided what to do next. This moment of hesitation is when you give the 'Stop' whistle/command, and it should sit facing you. A little spoken praise will reassure it that it is not in trouble, and it will wait for your next command.

Direct it clearly to one side or the other, preferably into the wind if there is any choice. Almost immediately it will scent a dummy, and will be immensely proud of bringing it back. Lots of praise is required so that it is full of confidence. Now repeat the exercise, and your dog will probably anticipate by returning diagonally to where it has just found its dummy. You don't need to check its initiative, but just wait until it gets to the spot it thinks will have the dummy. Being unable to find one it will hesitate, and you can tell it to 'Stop' as before. With its attention back on you, and its confidence restored by more praise, you can now redirect it to the opposite side. Because it is expecting to find a dummy in the central position it will probably be using its nose to locate one. If it is too confused you can always 'Stop' it again, before repeating the directional command. Eventually it will find the second dummy further to one side than it thought.

Very obviously it will get better with practice, and you can develop the exercise by putting other dummies beyond the first two in both directions. Now it will not be certain which way it is going to be sent, but will still believe that, by following your directions, it will succeed in finding several dummies.

Your new ability to redirect it can now be employed when you go for a walk. Dummies can be dropped along paths, or even 'tramways' in standing crops at right angles to your line of travel. Then, with your dog running ahead of you, a 'Stop' signal/command will bring it to the 'Sit' position, facing you. Leaving it there, retreat well past the turning along which you have secreted the dummy. Call it to you but repeat the 'Stop' so that it finishes more or less at the place where you want it to turn. The fact that you are going to redirect it from the 'Stop' position will be associated in its mind with looking for and retrieving a dummy – so it will respond keenly to your directional command/signal and will find the dummy. As you get more confident, and your dog more skilled and responsive, so you can increase the distances and complexity of this exercise. It really makes the dog listen to you and engage with what you want – much more than simple seen retrieves will do.

One disadvantage of teaching by Option 1 is that the dog will tend to stop too readily for visible barriers. Frequently, winged birds will drop on the far side of hedges, walls and fences, etc. A dog taught by Option 2 may also be reluctant to go through barriers, as it will never have been expected to do this before. Therefore, when you have achieved a good response to sending your dog in a straight line, you can start putting dummies (seen or blind) beyond gateways or gaps in the hedge. It will quickly learn. Eventually you will want to test your dog by throwing dummies into thicker cover – but that is a long way ahead yet.

An extension of this willingness to pass through gaps in a hedge is to train your dog to go through a gap, then stop it, before directing it along the far side of the hedge for blind retrieves. You can also throw a dummy over an impassable hedge and direct the dog along to a gateway or gap, where it will quickly whizz back on the opposite side before returning through the gateway with its 'prize'.

Water retrieves

Not all HPRs relish water. Most HPR breeds lack the true double coat of the Labrador and the other retrieving breeds, and feel the cold in consequence. Be prepared to be patient and understanding. Familiarity is everything and you should try to include some water experience from the outset. I like to get my pups used to hopping over little drainage ditches or small trickles. They invariably tiptoe around puddles to start off with, and I may have to lift them over streams. No matter, water is a constant presence in their world that is becoming ever more normal. As they get bolder, I expect them to splash through trickles in their need to keep up with me.

What to look for in a 'classroom'

Finding a suitable classroom for water training can be difficult. However, it is much better to delay the start of training than to try it in an unsuitable place. It does not take much imagination to realise that a gently sloping bank is the first requirement. This slope should continue under water so that the pup does not suddenly get out of its depth. This same facility needs to occur on the opposite bank. Very often this is too steep to allow a dog to climb out easily, and its subsequent struggles will tire and possibly frighten it. In advanced working tests and trials your dog will be asked to cross a river or pond, and it will do this more readily if it is confident of its ability to emerge easily. It is also important to have good footing, as slippery mud or slimy rocks can also be a deterrent.

Initially avoid fast-flowing water, and work mainly in ponds and lakes, although a gentle stream will not be too frightening. You should also inspect the classroom very thoroughly for hidden underwater hazards. Dogs can stake themselves on submerged posts, or get tangled in sunken fencing wire or other debris. Also make sure that the local youths have not been smashing bottles against rocks or concrete. For the first few lessons it is well worth avoiding any water that has wildfowl on it. There will be time enough to introduce this common source of distraction.

Introduction to water

For the first few lessons I would not recommend that you attempt to get the pup to actually swim. It is sufficient for it to feel confident enough to splash about in the shallows. It must be retrieving keenly, and you should be able to get it to run through shallow water without too much difficulty. A dummy dropped a metre away, in shin-deep water, is quite challenging enough to start with. It will also allow you to test the capacity of the retrieve article to float. Incidentally the traditional stick should not be thrown for dogs. There is a history of dogs getting them stuck down their throats when the stick lands with one end in the ground and the other pointing back towards the dog. Splinters can also be extremely dangerous and costly.

As your puppy gets bolder in the water you can ask it to go a little deeper. For most the sudden weightlessness they experience is agreeable, and they will happily paddle about. Others do panic and start flailing away with their heads held very high. Unfortunately, this can enable their hind feet to touch the bottom and, while this helps them get back to land, it stops them developing a low, efficient silhouette in the water. Only patience and going back a step or two will overcome this. Throwing the dog in really deep water is *not* the solution.

Eventually your dog should be willing to enter the water from quite steep banks, and to face fast currents. A real test of it, and of your training, is to be able to redirect it in the water. If you have an older dog that is good in water, it will give a lead and some healthy competition to a pup. Following it in will possibly take your youngster in deeper than it expected but, having survived the initial shock, it should want to rejoin its playmate. Without access to an 'aunty' you may have to adopt the role yourself, so dress appropriately (summer months with warmer water are definitely recommended in this instance). Be prepared to wade about (perhaps up to the waist) and get thoroughly soaked.

A lot of dogs will retrieve in water but then make for the opposite bank of the pond or stream. They know

Competition with older dogs will encourage puppies to be bolder

it is easier to run round than to swim back. Applaud their initiative – they are not being disobedient, just sensible. At an advanced level you should be able to send your dog across water for a blind retrieve of a hidden dummy on the far bank.

Tidal waters and spate rivers

Where you work your dog on coastal marshes and foreshores you must be wary of the possibility of it being swept away on a strong ebb tide. This danger can occur if you have a winged duck or goose that is still capable of swimming strongly. Your dog's efforts to retrieve it will only result in the pursuit going further and further out to sea. Eventually your exhausted dog will be unable to make its way back against the current. You may never find it.

Fast-flowing mountain rivers can quickly develop unsuspected and vicious volumes of spate water when there has been heavy rain further up their course. This phenomenon can occur so quickly that even a fit adult can be swept helplessly away. Many years ago, I lived by a quite small river in which a fisherman was lost in just this situation. Do not assume that your swimming dog can cope with everything!

Head shaking

Most dogs will shake their heads on emerging from the water in an attempt to get rid of water that has got into their ears. Too often they drop the dummy or bird to do this. Provided the game is dead this is not a serious issue for the practising sportsman or woman. However, if you have ambitions to compete in working tests or field trials, you will find that this habit can lose you points. Even if your dog is retrieving willingly and naturally, you may be forced to go back and teach 'Hold' as a command. Insisting on the article being retained will result in the dog either shaking its head while holding on to its burden or, more probably, waiting until after you have collected the retrieve article from it.

Retrieving over fences and walls

In the course of your dog's daily exercise it will have become used to scrambling through cover and over fallen branches. As you develop the retrieve it will equally learn to extricate the retrieve article from increasingly dense cover. However, bigger obstacles can be a problem. Firstly, the act of jumping and landing can put a strain on young joints, which can impair the dog's future level of activity. Therefore, do not attempt to develop this aspect of its work until its bones have really hardened off. In the larger breeds this can be well over a year old. That is quite a lot of lifting, hoisting, shoving and levering to do while it is young. But it is better to be safe than sorry.

Secondly, from a purely logistical point of view, it is not a good idea to teach an adolescent dog how to jump. Fencing your garden to a height that will defeat an active HPR can be quite costly, but a young dog that has learned to get out will test the efficiency of your fencing for the rest of its life. Conversely, a dog that has always been securely fenced is unlikely to question the impenetrability of even quite flimsy fencing.

Fences vary a great deal but the most common, and the nastiest, is sheep netting topped with a strand or two of barbed wire. This will effectively prevent the dog from getting through, leaving it to solve the problem of getting over, round or under it. If at all possible, the last two alternatives are much to be preferred, so teach your pup these skills from its earliest days in the hope that they will remain its first choices.

A safe technique for coping with barbed wire

Faced with having to get the dog to jump I can only advise you, wherever possible, to hold the dog back until you can lay your arm along the top strand. This will give it a clearly visible top line to aim at and, simultaneously, a firm surface to kick off. It is much better for it to 'bank' your arm like this than to try to clear it cleanly. In the latter case it may slip on take-off or misjudge the fence's height.

Because it is travelling with considerable thrust and all its bodyweight committed forward, there is the potential for serious barbed-wire tears at the very least. If you have a heavy dog it may be less painful to throw your coat over the top strand of wire to provide a safe route across the fence. There is also a very real danger with this type of fencing of the dog getting a hind leg trapped between two slack strands of barbed wire. The recommended solution is to flip the dog back over quickly in the way it came, and it will untangle itself. Trying to disentangle it while it is suspended is difficult and dangerous. The potential damage to its muscles, tendons, ligaments and joints is horrific.

These accidents can and will happen without the dog having to carry game as well. When this is the case it will instinctively calculate the extra effort needed to jump with a weight in its mouth, but often fails to realise that this burden may also get in the way. As a general rule follow the young dog and collect game from the take-off side of the fence, rather than expecting it to jump back with a bird in its mouth. Once you have taken the bird from it you can help it to get over by putting your arm or coat on the top strand again.

Personally, I think that if you cannot be bothered to help your dog over stock fencing, you should not have got one in the first place. By all means teach it to scramble over walls, through hedges and over (or under) gates, but be aware that jumping can cause severe injury and help make your vet a very wealthy person! The little pens used in gun-dog displays are to demonstrate blind retrieves and to prove that dogs can retrieve over fences. In practice it should only be limited to 'safe' fences, whatever they may be.

Runners

Regrettably a substantial percentage of game birds are not killed outright. Some may be so lightly injured that they will make good their escape, and will recover completely from their wounds. A lot will not be so lucky and will face a slow death if the gun and the pickers up do not make every effort to retrieve them and put them out of their misery. An experienced dog that will track and pick up runners is a valuable and respected addition to any beating line or stand of guns. Getting yours to that stage is a gradual process. Most dogs, on seeing a bird brought down, will anticipate a retrieve and will run out happily to collect it. But then they find their target is not dead. Should they point it or pick it up? Will they get in trouble for chasing it? How do they hold a struggling pheasant or duck? They may exhibit great reluctance to mouth a living creature, and will need every encouragement from you. Often a young dog will fail to do this for the first few occasions. If it has previous experience of retrieving canvas dummies garnished with pheasant wings, cold game and newly shot, warm

The dog should bring back injured game unharmed – it is your job to dispatch it humanely

birds it should want to make the attempt. Eventually it will pluck up courage to mouth at the bird. Some dogs, particularly soft-mouthed ones, take a long time to do this – so once again there are pluses and minuses in their behaviour. Hard-mouthed dogs will probably be more ready to grab a wounded bird – but may leave unnecessary teeth marks in it!

This situation is even more pronounced if the bird has been winged but is otherwise strong and active. Without the option of flying it will take to its heels in a fair imitation of the cartoon roadrunner. The dog you have been struggling to make steady is now being asked to course and catch live game. This requires them to snatch it up on the move, or pin it down to get a good hold on it. By and large, game birds do not fight back, but they will struggle to escape. It is a great achievement when your dog first delivers a wounded bird alive to your hand. Do not stint on your praise!

A close examination of the body, after you have humanely dispatched it, should reveal that its ribs have not been crushed by the dog's jaws but that there are indeed pellet wounds somewhere on its anatomy. For falconers the question does not arise – their quarry is either caught or it escapes.

The dog that will not retrieve

Some dogs are not keen on retrieving at all. This can crop up even if the dog has previously brought all manner of things back. It may have come to the conclusion that it is doing wrong because of some accidental mistiming of control on your part. You cannot always know.

The first thing is to leave well alone for a few weeks. Persisting will only serve to create an insurmountable obstacle in the minds of both of you. The more you persevere the more impossible the dog seems to find the exercise. Just get on with other aspects of the training. The plus side is that the less obsessed your dog is with bringing things back, the less likely it will be to become unsteady.

An alternative approach is to focus on the dog carrying things rather than retrieving them. Instead of trying to get it interested in chasing toys or dummies that you keep throwing for it, you can begin teaching it to hold easy, soft things. It may resist taking anything into its mouth and you may have to prise its jaws open, and then physically keep them closed on the toy. With lots of reassurance it will gradually accept holding most things.

Once this is achieved insist that it holds something before you put its lead on to take it for a walk. Every time it spits out the dummy, just put it back in gently with lots of reassuring praise. Only after the lead is on and it is still holding the object, can you take this back and go for the promised walk.

Another situation is if the dog is allowed to clamber on your lap or chair (the dog that does not retrieve is usually very sensitive and will benefit from little privileges like this). Only permit this if it is holding something. The same can be done with getting in the car, so that the dog comes to realise that holding things leads to good activities. It will begin anticipating this, and will bring a toy to you to initiate the desired reward. Praise will help develop this further until you can tell it to fetch something.

There is no reason why a young dog should not be expected to do some heelwork and stays, while it is holding a dummy. If you then move on to the recall, it is coincidentally doing a retrieve!

Introducing to gunfire

So far, we have not mentioned the use of guns at all. Until the pup associates them with game, they are just noisy, frightening things, and are best left out of the equation. Only when the pup is more mature and confident should this element of its work be introduced. All puppies are potentially gun-shy. For early experiences a starting pistol or .410 shotgun are ideal. If you have to use a twelve-bore use low-recoil, subsonic cartridges.

Up until now you will have been clapping your hands loudly when throwing the dummy, so the dog is unbothered by the noise and associates it with pleasure. The first lesson with gunfire should really be on a windy day, with the shooter placed a good distance downwind, so that the sound is carried away from the dog. He or she should be turned away from the dog, to shield the sound of the report. The shooter must wait until the dummy is actually being thrown, and the dog is therefore at its greatest point of concentration, before firing the gun. The dog handler should allow a slight delay, and only send the dog when its attention is back on the dummy.

Any sign of nervousness must be addressed by having the gun fired at a longer distance. You have to provide lots of praise to encourage the pup to concentrate on you – and the thrown dummy. Also get the shooter to make a fuss of the dog before, during and after the training exercises. This is an area where the use of food can be recommended as a training aid.

When your pup is unbothered by distant gunfire (100–150 metres away) you can slowly bring the shooter closer. If you have been using a .410 shotgun you can begin to introduce a twelve bore, with its louder report. The dog will start to associate the swing of the gun barrel with the arc of the thrown dummy, something you can introduce with a stick at an earlier date.

Remember that a dog with sensitive hearing needs more care in training to the sound of a gun, but will also be much easier to control by voice or whistle. The reverse also holds true – a dog that is unbothered by a gunshot will possibly not react too promptly to your audible signals either. As in all dog-training dilemmas there are pluses and minuses, not right and wrong. Your skill is to maximise your dog's good points and minimise the bad ones. You may not succeed in having a perfect dog but, by doing this, you can have a perfectly successful one.

Problem solving chart

Topic	Problem	Solution/s
Retrieve	Lacks interest in retrieving	Use its preferred toys Incorporate toy in play without throwing it until it is sufficiently interested Ensure toy has weight and texture that is suitable for puppy. You have been too ambitious – go back a step or two
	Reluctant to release article	Delay taking article until praised sufficiently Give food reward once pup returns with article Use less attractive article
	Dropping the article during the retrieve or on its return	Back off quickly to keep its attention on you and to speed up its return Practise carrying article at other times Introduce extra articles during retrieves
	Mouthing the article	Increase the weight of article and speed the return up as above Practise carrying article at other times
Cover	Reluctant to enter cover	Gradually introduce progressively deeper/tougher cover Increased enthusiasm for retrieve Toughen pup physically with rough play
Obedience	Will not wait for retrieve command Runs off with article	Do short retrieves on lead Collect article yourself Develop basic obedience more Walk away, make it follow Practise carrying article while doing recall on lead Allow it to carry article before taking it Increase weight of article
	Chews or mouths article	Increase weight of article Change to different texture Increase willingness to return

Chasing	Anticipates you throwing article	Improve obedience particularly by tempting it with kicked or thrown articles while it is sitting Pocket article and practise recall and stays Drop article without dog seeing and subsequently send dog to search for it
Water	Reluctant to enter water	Choose easy access and very shallow 'splashes' Cross water yourself to enforce following Enter water yourself and give food reward Involve other dogs Have very buoyant article for visibility Go back a step or two
Game	Reluctant to pick up game	Tie feathers/fur or wings to favourite articles Accustom to cold game initially Generate play with cold game/feathers/wings See Obedience (above) 'Runs off with article'
	Reluctant to bring game back	Encourage carrying while doing heelwork (distinguishes carrying from chasing/killing)
	Mouths or chews game Very hard-mouthed	Wrap fur or feathers round hard, heavy article, e.g. small log Cover plastic water bottle with fur or feather, and fill with distasteful substance/liquid, e.g. vinegar, for less attractive article

Chapter 6
Hunting

Hunting is the most natural thing for an HPR to do. It consists of covering large areas of ground to locate potential quarry. The wolf, coyote and fox still do this of necessity and use their eyes, ears and nose for the job. Apart from needing the physical equipment with which to hunt, your dog must also ***want*** to find game. Only time will tell if your puppy possesses this overwhelming urge – but generations of knowledgeable breeders, and selected ancestors, are on your side. You can develop it by taking it to game rich-areas, and really encouraging every little investigation it makes. This can begin from its earliest free runs.

Scent

Each sense can be inhibited from working to its fullest by ambient factors in the environment. Equally some conditions are more conducive to success. With vision we know that sight is impaired increasingly as light fades, until an animal can no longer see in total darkness. If the light is behind the hunter it gives the best illumination possible, while conversely its quarry will have difficulty looking directly into the sun. Equally our hearing will be less efficient in a noisy environment, and it is much harder to hear sounds that come from a downwind direction. In quiet, still conditions our hearing can be very acute.

Pointing dogs live in a world at which we can only guess. We can never share their experience, or understand the richness and diversity of information that is carried to them on the wind. Despite this we need to grasp something of how scent works.

How scent works

Scent consists of minute quantities of substance which have 'broken away' from the source. The size of these tiny molecules is such that they may be light, or volatile enough, to be carried long distances through the air, and these give us '**air scent**' or '**wind scent**'. Alternatively, they may be so dense that they rapidly fall to the ground, or attach to surrounding material. This creates '**ground scent**' or '**foot scent**'. These larger, heavier particles will have a stronger odour, and will hold this for longer before the scent deteriorates.

Air or wind scent

These terms are simply two different names for the same thing. They occur when tiny molecules of scent vaporise, or are released and then carried and dispersed by another more volatile gas. Typically, this might happen where heat causes vaporisation. The smell of food cooking is one we can all understand, as is the first airborne hint of brine that tells us we are getting near the sea.

How far can air scent travel?

This depends on three things. Firstly, the amount of heat that the source of the smell is generating will make the scent rise. Secondly, the strength of the wind will dictate the distance which scent can travel. Lastly, the scent will survive longer if it is attached to a more volatile substance that will itself last longer and can therefore travel further.

If I put some sage leaves (a fairly pungent herb) into a stew there will be only a moderate amount of scent until I apply heat. As the stew starts to cook, the water content begins to steam, and all the molecules of scent from all the ingredients 'hitch a ride' on the water droplets. I will even be able to identify the different ingredients. As the steam rises it may get carried out of the kitchen window and be picked up by the wind. The warmer the molecules are, in comparison to the outside air temperature, the more rapidly and higher they will rise. Wind moves much more quickly at a greater height, so this helps disperse the smell of my stew even further.

Those smells that we think would mask odours may actually carry them. The more volatile the substance the better they will carry scent. Remarkably it has been found that drug-detecting dogs are more successful when narcotics are hidden in smelly places like petrol tanks, because the vapour is more volatile.

Understanding the importance of wind direction

Because it depends on vapour to carry it any distance, air scent is totally governed by the wind's direction. It cannot travel against the wind at all. In practical terms this means that your dog can only detect air scent from sources that are upwind of it. When the dog is working into or across the wind, the oncoming breeze brings it early warning of everything that lies ahead. By contrast, when the wind is coming from behind it, the dog will reach the source before it discerns the air scent, as this is being continually carried further away and is being dispersed. Knowledge of this factor is basic to your understanding of working and training a pointing dog.

If you think of a smoking bonfire you will get a clear idea of how scent is dispersed. As soon as the smoke emerges it is blown downwind. A strong breeze will quickly lift and carry the smoke away, and it will become less visible to the eye, until even the faintest haze is no longer detectable.

On a still day, in contrast, the smoke will slowly climb and get thinner, but will not travel over a wide area. In cold weather we can sometimes see a cloud of smoke hanging above buildings and failing to disperse at all. It does this when the original heat cools down and ceases to offer any lift, and when there is no wind speed to move it away. On hot days the cloud of smoke keeps rising until it can no longer be seen.

This principle dictates how scent works, although it is not visible to the eye. It is most easily detected where it is at its greatest density, close to the original source. If there is a breeze blowing the scent disperses quickly. By turning into the wind and towards the scent which caught its attention, a dog will immediately enjoy an increased concentration of scent, until it eventually arrives at the source. Turning the other way, with the wind behind it, results in an instantly identifiable reduction in the amount of scent available. On still days the scent does not travel as far but will remain strongest close to the source. The dog will have to cast about to find the direction that gives the most satisfying result, and rapidly decides which way to go. It will do this so quickly that we are often unaware of any hesitation, but it will have had a moment of uncertainty. This ability to distinguish thickening scent from fading scent also enables dogs that are working on foot scent to follow tracks of animals or people in the correct direction. Any mistakes are rectified in seconds.

Working your dog into the wind gives it the best chance of finding game. Real life is not so simple. Wind has a habit of swirling around and changing direction. On open ground this is fairly minimal, but in enclosed country every tree, bush, hedge, building or hillock creates eddies of wind, and consequently disrupts scent. The pointing dog takes a fairly generalist approach to this and, because it is searching for wind scent, is usually some way off its quarry when it first hits scent. You will find as often as not that your dog is as likely to 'tack' slightly into the scent as it is to work straight into the source.

What lifts scent is the contrast in the temperature of the scent (or what is carrying it) and that of the surrounding air. In freezing conditions, the warm breath of a bird or animal rises quickly. However, because water droplets cool rapidly to the ambient temperature, it will soon lose its potency unless the wind is pretty brisk.

Once, in bitter sub-zero conditions in the Scottish Highlands, we were working some snow-covered water meadows. These lay along the bottom of the glen and were bisected at intervals by occasional strips of low plantations which were destined, eventually, to become windbreaks. My bitch suddenly checked while quartering across the prevailing wind and turned into it. At a steady but determined pace she trotted the length of the field we were in until she reached the windbreak. Unhesitatingly she popped over the rabbit fencing that surrounded it, trotted on through the young trees and out over the far fence. Without altering her stride, she carried on across the length of the next field in an unwavering line. The next windbreak was treated in the selfsame fashion until, just beyond the second plantation, she went on solid point on a rabbit that was crouched down beneath a tussock of grass. We calculated that she had picked up the scent some four hundred metres away.

What was significant was that the scent had been above ground and, while she must have passed half a dozen occupied warrens on her way, this was fresh, warm scent that was carried to her quickly before it could cool down.

On a frosty day you could, therefore, reasonably expect any dog to locate an occupied rabbit warren. The contrast between the surrounding cold air and the warm breath of living things must be immense. However, since you can probably see for yourself where the grass has thawed around the mouth of the hole, this is really not much of an achievement. Finding shot game is also easy as it cools slowly and does not stop releasing warm scent for several hours. By contrast cold game, which you might use for training or in competition, is already at the same temperature as the air and the scent is less volatile. Try smelling both fresh and cold game yourself, and then compare them with similar creatures direct from the freezer, and you will readily appreciate the difference.

When the dog's nose is least effective is when the wind is behind it, which is a direction it will show no preference for. If it stumbles across quarry, and there is no other description for when this happens, it will almost tread on the game at the same time as it finds the scent. You may curse it for having failed to point your rapidly disappearing quarry when, in truth, it never had a chance.

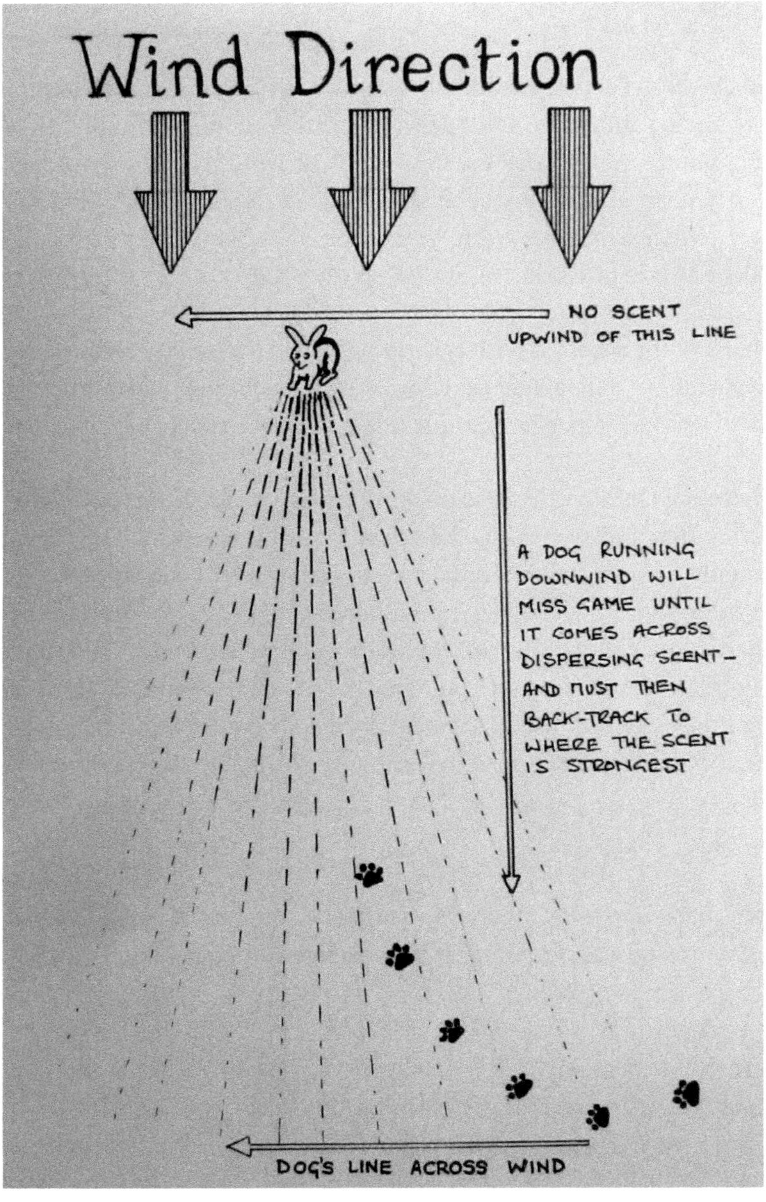

At other times, when working downwind, it will overtake dispersing scent from something it has already passed. It has only caught this as the scent has spread wider and the edge of it has extended into the dog's path. If you watch closely it will work back towards the wind, but to one side of its original line of travel.

A fascinating example of the dog's ability to understand scent occurs when you see game get up, without it having been seen by the dog. Within seconds the dog may run across the cooling scent from the spot where the game was sitting. Yet an experienced dog will not even check at this, but will quarter on without a break in its stride. It must be aware of residual warm scent but can instantly discriminate between this and the real thing given off by the live animal. Inexperienced dogs will pause and briefly examine the place, and a dog which false-points may give every indication that game is still there.

The dog's nose is claimed to be approximately ten thousand times more sensitive than ours. It is able to detect the tiniest traces of scent because of the number of scent receptors in its nose. If we could spread our own receptors out, they would cover an A4 sheet of paper; doing the same with a dog would cover a football pitch. We know that warm scent at body temperature is the most easily detected on the wind and that this is less easy with cold scent. The latter, in turn, tends to stick on the ground or vegetation and provides a more permanent but less volatile scent. To maximise the dog's potential and to encourage it to seek air scent, we usually try to work it into the wind.

Foot or ground scent

Particles of moisture, when they fall, deposit minute traces of the scent they have been transporting on to the ground, where they stick to the vegetation, soil and other surfaces. The heaviest particles, which are the ones most saturated with scent, will fall first and these will also continue to emanate for the longest time when on the ground. Even in virtually arid conditions some scent found close to the ground will remain, and is very difficult for the dog to detect. What remains is known as **foot scent**. Its presence is much more long-lived than air scent which, by its very nature, disperses. Bloodhounds and foxhounds typically work close to the ground to unravel complicated trails, but then switch to air-scenting as they near their quarry. All dogs can distinguish the scent of their quarry from that of other animals, or of trampled or crushed vegetation.

A friend of mine saw an Egyptian police dog demonstrate just how acute the dog's sense of smell can be. In the centre of Cairo, in intense heat with virtually nil humidity, he watched a trained dog successfully follow a 24-hour-old trail through a busy market place. Dogs imported from more humid countries took a while to adapt but quite quickly learned to cope with these conditions.

Quartering

This term does not mean that the dog works in one quarter of a field at a time, but relates to the systematic covering of ground. It occurs when the dog works in straight lines, crossing backwards and forwards at right angles to the wind and, coincidentally, to your own line of travel. At its best it can be metronomic, with the dog zigzagging from left to right in front of you. The extent of its outrun on either side should be dictated by you and, in an ideal world, should be equal and balanced. You can possibly create this behaviour in your dog by habit, irrespective of the wind direction, but this requires obedience rather than initiative. In a pointing dog it should always be as a response to the prevailing breeze, so you must get into the habit of using wind direction from the outset.

At fourteen-weeks this puppy is already very attentive

Puppies have, or at any rate show, little awareness of the wind. They explore fairly randomly although they will mostly investigate scent which is upwind of them, for the simple reason that they are ignorant of any scents that are downwind. It is natural for them to range out a short distance before turning back to the known, and safe, place from which they started.

This is how fox and wolf cubs develop. As they become more active, they will begin joining their elders and tagging along. Your pup effectively does the same when you start taking it for walks. At first you are only interested in teaching the pup to keep an eye on you, and to respond to your call. As it gains confidence and strength, and you learn to trust its response better, you can happily allow it more freedom. You will also, I hope, have got it used to expecting you to disappear in the opposite direction whenever its attention is distracted.

By the time it is about four or five months old you can begin to organise this behaviour into some semblance of quartering. It will only be a brief introduction and should not be seen as an attempt to create a quartering robot.

Choosing a "classroom"

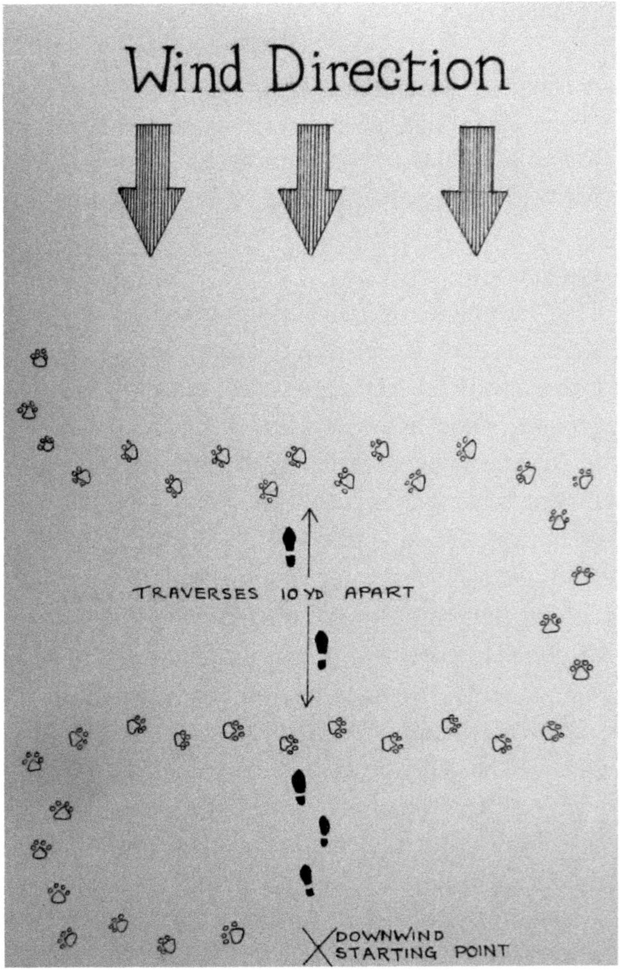

To encourage quartering as a purely artificial exercise you should choose a suitable area. This should be open grassland with some natural scent about to keep the pup's interest, but without much probability of actually finding game. You do not need big fields for this, in fact a relatively small paddock is preferable. Quite frankly I would struggle to use anything much bigger than a football pitch. My reason for suggesting this is that you should, from the outset, encourage the puppy to work to the edges of the field. It helps if the length of the field runs along the line of the prevailing wind.

Having selected your classroom, wait until you have a mild day with easy scenting conditions and a steady wind from the right direction! You want the wind coming down the length of the paddock. Take the pup, on the long line, to the centre of the downwind end of the paddock so it can only go into or across the wind. A couple of stay exercises will get its attention on you before you start.

Imagine that you start your pup running off to the left so that the wind is hitting its right cheek. Use a hand signal and lots of verbal praise, and run after it until it gets up to the fence. Just as it reaches this you must turn back to your right and give a short, sharp, single '*peep*' on the whistle. You are now running towards the opposite side and the pup should hurry to follow you with the wind now on its left cheek. Encourage it to pass you and continue to the paddock boundary, where you repeat the process back to the left again.

Because the puppy is excited by your activity and enthusiasm (and it is essential that you do generate this), it will be racing across in front of you (and probably jumping up and tripping you over as well). Each zigzag will take you further up the field until you reach the far end, by which time you will be thoroughly exhausted and your pup will be overexcited. Quite possibly you will only have done ten traverses, with each one lasting some fifty metres for the pup and a bit less for you. When did you last run four hundred metres? After a good rest, return to the start point and, following another couple of stays, repeat the exercise.

So far you have got the pup to turn on the whistle and to zigzag across the field. It has been fun but nothing more. If your pup is not responsive you can use a light, long line or a retractable lead. In this case, from the very first lesson, give a sharp jerk on the lead at the end of each traverse, as you change direction. At the same time give a single, sharp '*peep*' on the whistle, and it should start to respond quickly to the whistle to avoid the jerk. Provide good praise to encourage a prompt and willing response in the future.

A reason to quarter

After just a day or two of this the puppy will start anticipating the end of each traverse, and will be turning short. You do not want this so, as it turns short, you can stop the pup and redirect it towards the fence again. Throw a morsel of smelly food to the base of the fence with an exaggerated arm signal in that direction. This encourages the pup to go right up to the fence on each outrun, even though the use of a reward should be random and infrequent.

If you are lucky the pup may offer you the chance to develop its nose a little more. As it checks and starts to 'foot-scent' along the fence, you may be able to flick another reward (or possibly a dummy) further along the line in which it is travelling. Because this is upwind of it, the scent will help it find the food. This can be a desirable habit, and I like to see a dog work right up to the boundary of its ground, then follow the boundary for between five and ten metres before turning back across you again. Almost certainly ten metres between traverses is much less than the dog can cope with, but you will be more confident yourself if you feel it has covered its ground thoroughly.

Some purists may be unhappy at the thought of encouraging a pointing dog to work fences, and particularly hedges and ditches. They may also shudder at the idea of encouraging the dog to look for ground scent in this situation, as this is not the pointing dog's speciality. However, a lot of sportsmen and women hunt countryside edged with hedges, dykes and walls. Very often these holds game and should not be overlooked. You should also remember that the wind can swirl around this kind of cover, and you may need the dog to check it a little more closely.

If you are lucky enough to work grouse moors exclusively you may never need this skill, but you will still value a dog that runs out purposefully, until it is turned by your whistle. Although you and your dog recognise the boundaries of the land you are working, you must remember that the game may not. Sometimes, by way of variety, get your dog to work through hedges and include the opposite side or jump across walls to take in some of the adjoining field.

Years ago, I ran a course for novice HPR owners. This included quartering a fairly narrow strip of field, flanked by an overgrown hedge on one side, and an equally overgrown ditch on the other. They were told to encourage their dogs to hunt out to each boundary. Through self-consciousness few of them generated any excitement, and therefore their dogs did not show sufficient 'outrun'. The rest stood 'coffee-housing', paying no attention to the success of their fellows.

After they had all tried, I took my old dog over the same ground. He covered it with his trademark zeal. Halfway along he took a solid point, where a bramble bush overhung the ditch. I called one owner, a falconer, to bring his promising pup over. He ambled up, his dog pulling this way and that on the lead. Eventually, with his half-hearted encouragement, the pup picked up the scent and was allowed to flush a rabbit. Its whole attitude changed immediately.

Some while later I asked how his pup was working. 'Oh, I don't bother taking him any longer. There's nothing about and he just spends all his time with his head down rabbit holes.' When I suggested that the addition of a ferret to his team might yield quite spectacular results, I was met with a blank stare of incomprehension.

To hunt successfully both your dog *and* you must want to catch game. Keen hunters have keen hunting dogs! Sadly, few of the other handlers, who were watching that night, took on board that they too had failed just as badly to motivate their dogs to find this rabbit.

Leaving the classroom

The small fenced paddock soon becomes too small and restrictive. How long you use it for will vary from pup to pup, but no more than three or four sessions will probably be enough to give you both a basis for quartering.

As soon as my pups start responding to the turn whistle and begin using their noses, I like to get them out in the larger world. There will never be an adequate substitute for the real thing, and you and the pup need to develop initiative and some mutual trust. This does not mean that you should not take advantage of any situation that resembles the classroom, i.e. a defined area with a steady headwind, to have a little practice. You should still encourage the pup from side to side and hope that with the attraction of new smells its outrun will gradually increase, so that you do not need to run quite so far with it on each 'tack'. Particularly get it interested in any isolated island of cover, such as a bramble patch or a clump of thistles. Call it over to these and try to get it to approach the area from the downwind side. Places like this will probably hold some scent, and teaching your dog to look for them will add to its skills.

After the initial lessons in quartering I rarely return to the classroom for the remainder of the first season. Only in the following summer when the dog is stronger and more switched on to game, do I resume quartering as an obedience exercise. Even then I will use much bigger fields, and will consciously walk directly into the breeze to ensure the dog is working at right angles to me.

This is also an opportunity to establish an equal 'outrun' on either side. If your dog runs wider to the left than the right it can be worth walking on a line that is left of centre, so that the dog has to run further to reach the right-hand boundary. The reverse is done where the dog's preference is to go wider to the right.

Wind direction

By now you should understand that your dog can only find scents that are being brought to it on the wind. This knowledge will enable you to encourage hunting/quartering behaviour to maximise the amount of scent your dog can pick up. We speak of three basic categories of wind direction:

- **headwind** is when the dog is travelling directly into the wind. Any scent it picks up will be pretty much straight ahead of it, and it will be tempted to steer towards the source in a fairly straight line. This is inefficient, as it would only be covering a very narrow strip of ground, and dogs that do this will tend to run ahead of you, rather than quartering across your line of travel. Often this is the result of you regularly taking 'linear' walks that go to specific destinations, instead of using the opportunity to work the dog across a broader front. If your dog will quarter to your hand signals you can encourage it to work over a wider front, as long as it is prepared to ignore any distant scent that is ahead of it, and in a competitive situation you may be forced to do this. The challenge will always be to hold it back.
- **backwind** is the exact opposite to head wind and will be coming from behind you. The dog will only find scent once it has gone past the source of it, and has strayed into the widening fan of scent as this spreads and diminishes. When this occurs, the dog will turn back into the wind and work back towards the source of the scent. A backwind can also cause the dog to flush game without pointing – as it has literally stumbled upon it before reaching the scent. If you find yourself travelling with the wind at your back you will need to keep turning the dog across your path of travel so that it runs across the wind. Progress is normally quite slow in this case as the traverses are compressed closer together by the dog continually checking back on scents.
- **cheekwind** is what you always aspire to your dog using. It is the product of the dog running across the head or backwind so that it is sideways on to the wind direction. This means it will pick up all the scents that lie upwind of it, and the further it runs across you on each traverse so the amount of upwind land it covers is increased. Where this is problematic is when the wind is at right angles to your intended line of travel. For the dog to keep this on its cheek it must run forwards and backwards in the direction you are walking in.

The ideal quartering pattern

Novice HPR owners often want guidance about the width of outrun, and the space between each traverse. Both of these measurements are variable, which is little help until you learn to recognise that these matters are dictated by the amount of cover, the dog's nose, the wind speed and the scent. Always remember that it is the wind direction and not the shape of the fields that determines the dog's (and your) line of progress.

In an oblong field with the wind coming straight down its length, you would walk centrally, directly into the wind. Meanwhile the dog should be crossing in front of you (so that you do not flush any game yourself) as it traverses from one side to the other. By doing this any game in the field which is upwind of it will be found. This is the same kind of pattern you taught your dog on its first lessons.

In the same field, but with a diagonal wind direction (see diagram), you would be advised to walk it diagonally so that you are at right angles to the dog's traverses across the wind. A perfectly straight course into the wind would give an unequal outrun for the dog, unless you deliberately weave in your course so that you are always about midway between the two sides. The dog that works properly to its edges will do so irrespective of your position in the field.

As you walk into the wind, and your dog quarters at right angles to it, it is maintaining a cheekwind. This is one that blows on to the side of the dog's face. The wind brings scent to it from the land that lies ahead. I think you should picture your dog 'harvesting' the wind as it works to and fro.

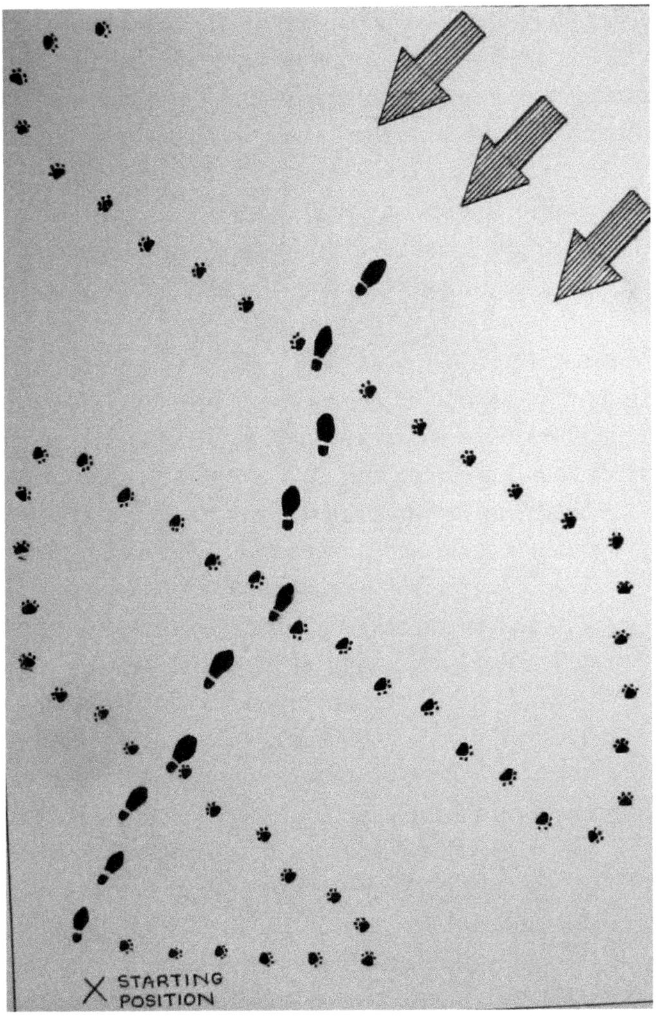

Fields are rarely perfectly rectangular, so try to walk directly into the wind, and midway between either side.

Left to its own devices the frequency of the dog's traverses, and the gap between them, will actually vary according to the wind speed and the temperature. Do not think that the dog is working any less thoroughly because the gap between the traverses is greater than the last time you had it out. Equally, working a much tighter pattern does not necessarily mean that your training is finally having an effect. More probably it is just struggling to find scent.

Working speed

A young dog with plenty of drive will, once it gets keen on hunting, begin to work with great enthusiasm. It may cover the land at high speed – and with wide spaces between each cast. You may feel it is getting out of control and is rioting, and that you should start working to control the whole pace of the exercise. This may prove a pretty futile attempt. Instead, get inside its head and understand that above all else it wants to find game – and as soon as possible.

It will have greater faith in its own ability to detect scent than you do, and is simply hoovering up as much as it can. **It will not outrun its nose** – although with experience it may become more careful or thorough in adverse conditions. I have come to believe increasingly that a fast dog has probably just got a better nose than a slow one! Does an older dog become more thorough (or just more economical with their energy?) or is it that their noses become less sensitive as they age? I have never heard of anyone researching this.

(Photo by Andy Coveney)

This young dog is quartering fast, but the slight tilt of its head shows that it is 'harvesting' the wind as it runs, which is coming from its right.

I have just finished working a young dog (sixteen months old – see above) during his first season. He has tremendous drive, and will work all day at a frighteningly flat-out gallop over a very wide front. Happily, our gamekeeper understands that he is still working and not rioting. I also work the dog's mother, often in tandem with him since he shows no interest in following her. To be truthful, in the very heavy cover on this shoot, I just find it reassuring to have at least one dog somewhere that I can see! She has a happy knack of finding him when he is on point, and backing his point – which is a great help to me.

Although they are working the same ground, and she is a seriously experienced and reliable game-finder, she rarely finds anything he has missed. His speed does not make him inefficient in any way. Neither is he being inattentive. I know this because he put up a hare the other day, when he was some 300 metres away from me. I gave a single recall whistle and he spun round and shot back to me immediately at the same gallop at which he does everything else!

Working cover

It is often impossible for a dog to work a precise pattern in heavy cover. A favourite hillside of mine is dotted irregularly with thick patches of gorse. My dog will not hold a straight line, but will skirt round these, preferably to the downwind side. By doing this he can tell if anything is harbouring there. If there is, he should penetrate the cover until he has located his quarry; if not I expect him to run on in the same general direction to the next piece of gorse. In practice he will normally circle cover that holds game until he has pinpointed it from the outside, and has identified the shortest route to it. I am sure he does this to minimise unnecessary prickles. My concern is solely that he covers the hillside thoroughly, not with maintaining any visual pattern. In this instance, because I have very good visibility, I am more than happy to allow my dog to run perhaps as much as one hundred metres to either side.

We also work the kind of woodland that is typical of much of the British countryside, i.e. mixed deciduous woodland with bramble and bracken undergrowth. Among the bigger oaks and ashes there is a median layer of birch, hawthorn, hazel and elderberry scrub. Here visibility is such that I want the dog to work much closer to me. Because his early training taught him to always keep one eye on me, he automatically shortens his outrun and reappears at regular intervals. We are not telepathic but I have only to think '*Where is he'?* and he turns up. His ingrained need to keep in touch, which I taught him and obviously share, has programmed me too.

With my first HPR I got quite cross that she showed no enthusiasm for entering thick cover. Her preference was to follow the footpaths and tracks that threaded through the bramble undergrowth. I had expected her to behave like the average spaniel, to whom pain seems a desirable thing in itself. She taught me that her nose was far superior, and that she would readily plunge into the thickest briar patch – if there was good reason! Because HPRs 'read' the land, in a kind of shorthand, they cover much more ground, more quickly but just as thoroughly as conventional gun-dog breeds. Much of the difficulty the novice HPR owner has is simply down to our lack of trust in the ability and focus of our dogs. We need to realise that they are actually always working, and never messing about.

When we work very open ground, such as moorland or hill country, I do not need to worry about his whereabouts, because I can see over much greater distances. Here it is quite usual for him to work some two hundred metres away on either side, because he is able to keep a check on me as well.

What is the ideal distance between traverses?

Ideally the dog should traverse (quarter) in front of you, so that it never covers the same ground twice, and equally has not missed any part of that ground. In good scenting conditions, on open land, this may mean each traverse is done forty metres or more beyond the previous one. When the wind is blowing hard this may become much less, which is also the case if there is much holding cover on the ground. Scent may persist here, requiring more thorough investigation.

A partnership of handler and dog often arrives at its own quartering arrangement. I find that I am comfortable if my dog covers the ground so that I can walk at my normal pace. For example, if he covers a one hundred-metre front, with traverses forty metres apart, I will walk forty metres while he runs one hundred. Should he work a tighter pattern, say an eighty-metre outrun with a traverse every thirty metres, my pace is virtually unaltered.

I will have to adjust my pace when the dog works a very wide front in poor scenting conditions. Because he needs to quarter more regularly, I find in this situation that I am waiting for him. This is not a hardship on a mild day but, when the wind is howling across Perthshire and straight through every supposedly insulated item of my clothing, I have been known to shorten his outrun so that we can move on more briskly and my circulation may recover!

Working across and downwind

Working your dog into the wind makes it easy to develop a regular, dependable quartering pattern. This is highly desirable and gives you confidence that your dog has covered the ground thoroughly. It can be so efficient that you can denude an area of game, simply because you quickly find all that is there. If your job is grouse counting then this is what you want, returning to a fresh starting point at the downwind boundary of each designated 'beat' or parcel of land.

By contrast, the rough shooter or falconer has an interest in maintaining a stock of game on the land. There is no need to be 100% efficient. What is missed will be there for another day, and will form the nucleus of next season's population. So, it is often sensible to work in a less regimented manner.

Imagine you and I have gone out for the morning along a river valley. It makes sense to follow the meanderings of the river where we know that reed beds along the bank are likely to harbour game. Any pastureland that flanks the river will be grazed very short, and there will be little (but always some) chance of finding quarry on the open fields. As a result we concentrate on the waterside, getting the dog to work more like a spaniel.

A little further on, the river swings in a ninety-degree bend, and we find the wind is no longer in our faces. The next field is sown with kale, which may hold a few pheasants from a neighbouring shoot. We could leave the bank and walk down the fence furthest from the river to cover this land in the accepted and more profitable direction, i.e. into the wind. Otherwise we can keep to the river and let the dog range out across the field as we go. With a young dog I would definitely recommend the first choice to maintain its good habits, but there is an interest in watching an older dog demonstrate its skills in less favourable conditions.

For a falconer or hawker it is well-nigh impossible to cover countryside methodically. Any slip after game can take you in a random direction, and it would be stupid to return to the spot from which the game was flushed. By the time you have relieved your bird of its victim, the dog will have recommenced operations from this new starting point anyway. You might just as well continue from where you are. I find that no matter what Plan A was at the start of the day, our actual progress will be haphazard. Nevertheless, we do cover the ground pretty well in a rather 'freehand' manner, and the dogs learn to work downwind perfectly efficiently. This has taught me that competition-style quartering is not necessary to the successful outcome of the hunt, although it is pleasing to the eye.

By contrast a 'walked-up' day on partridge, where your dog is replacing a line of beaters, should be done methodically. Provided you can choose the best direction to walk, i.e. straight into the wind, and provided that the dog handler is central to the drive, then any dog with a good outrun and the training to stop and turn on command, should quarter in textbook fashion.

Many years ago, I used to exercise my dogs on some common land, which consisted of a large expanse of rough grassland. On most days I would see an English Setter quartering this in search of skylarks, with which it was obsessed. Apparently, its owners simply let it out for hours at a time, confident that they knew exactly what it was doing and where it was doing it. This self-taught dog quartered very stylishly indeed.

I have come to the conclusion that the dogs know more about the job than we do. They never fail to amaze me with the extra bits they introduce of their own accord. Some will, while on point, run quickly to the opposite side of the cover so that the quarry is boxed between you and them. One bitch of mine would, if on an undetected point in heavy cover, wait for perhaps five or ten minutes before becoming so frustrated she would take matters into her own hands and flush the game. Often this would be a rabbit and a volley of hysterical yelps would precede her as she pursued her quarry. This resulted in some very strong language but she never got out of the habit. After a regrettably long time, I became intrigued that the rabbit always seemed to be heading directly towards me. My hawks picked up on this much sooner and would, on hearing the noise, look for the rabbit's arrival. It slowly dawned on me that she was herding the game to me, rather than just chasing it. This was confirmed on a couple of occasions when the rabbit, on seeing me and the hawks, jinked off its line. She at once lost interest and switched to quartering silently again.

Fairly early on in your partnership with your HPR, it is a good idea to allow the dog to work as it thinks fit, for perhaps an hour or so. There may be no recognisable pattern to its work, and you may not think it has done a good job. Then take it back over the same ground and really work it thoroughly, double-checking every section and corner that it seemed to ignore the first time. Almost certainly you will find that it has missed nothing. You will gain a lot of confidence in its skills, and respect for its abilities, by doing this.

The same bitch, who was never taught to retrieve and showed no interest in this activity at all, put us to shame one day. I had gone out with some clients who had their pet HPR with them. They were anxious to see if it would work, so I kept my bitch on the lead. Their dog showed little interest, resolutely refusing to point anything, and the day was not going well. In the end they agreed to let her work, I think because they suspected there was no game about. As soon as she was let off, she disappeared into a small reed bed that their dog had just quartered, and reappeared with a live rabbit in her mouth. Having delivered it perfectly to hand she never did another retrieve in her life.

Scent Shadow

This is a seldom-mentioned aspect of wind scent, and occurs where the wind is coming through a fairly high hedge, or round any other large obstruction. As the dog nears the hedge it will hit both ground scent at the hedge bottom, and air scent as it filters through the branches. Your dog, instead of moving on, does a 'double take' and checks back along the hedge. This is sometimes frowned upon on the basis that the dog should have already covered this ground. However, what the dog a second wave of air scent which has billowed over the hedge, with more information. I think I would prefer a dog that makes sure over one that works in a textbook fashion.

Fault	**Cause**	**Remedy**
Dog shows no interest in hunting	Lack of game scent	Get access to stocked land Use warm game as retrieve article Practise blind retrieves so dog has to hunt for the retrieve article
	Has been overcorrected for chasing or killing	Reduce level of control
Dog is reluctant to range out	Overcontrolled Lacks confidence	Reduce excessive control Allow dog more opportunities to be wayward Follow dog to remove its insecurity, instead of moving away from it
	Over-interested in you	Reduce praise Work between two people to break dependency on just one
	Lacks incentive	Use warm game or caged game
Dog does not use wind	Lack of game scent	Get access to stocked land Use warm game as retrieve article Concentrate more on blind retrieves
	Wind may be swirling so no pattern is possible	Remember to check wind direction continually
Dog does not work tight pattern	Light or no wind	Practise on more suitable day
	Wind is behind dog	Work into the wind
	No game scent to focus dog's efforts	Use scent-holding land
Dog reluctant to enter cover	Physically sensitive	Reduce sensitivity by boisterous handling
	Thorns, nettles, etc.	Accustom to soft cover first Accept dog's reluctance as sign of intelligence It will enter cover holding game Use retrieve as incentive

Chapter 7
Pointing

For the purpose of this book, I have taken pointing to include the whole process, from the dog having first found game, then pointing it, permitting your approach, waiting for the command to flush and, finally, ignoring the temptation to chase the flushed game. Anyone who has worked HPRs will immediately spot the many pitfalls between this casual résumé and reality, when all or any of these facets can rapidly disappear from the dog's repertoire.

At the start we looked at the dilemma between obedience and initiative. These twin pressures are dictated by the individual dog's temperament and our own attitude. With any sort of scent work you are wholly dependent on the dog's perception of an olfactory landscape which we can never appreciate. We have to trust the dog because it knows better than us. This is why you must concentrate on its obedience at all other times, so that you know it is probably working and not just messing about. Happily, the strength of breeding is such that most HPRs are virtually unable to play when out. All they do is work. Once you have come to realise that you are part of a team, rather than in a master/servant relationship, you will do much better.

When I got my first HPR I looked everywhere for books telling me how to teach my dog to point. Although I searched through any number, I could not find one that set this out. This puzzled me greatly until my eight-week-old pup started pointing flies in the backyard. What this taught me was that the pointing instinct is bred into the dog and is not really something you can teach as a set exercise.

Since then I have had the good fortune to work a fair number of HPRs. Of these only one did not point at all. I kept it for twelve months in the hope that my perseverance would finally achieve some sort of breakthrough. Eventually I did what I should have done sooner, and reached a decision to find it a pet home. Before this happened, I was unfortunately persuaded, against my better judgement, to let a friend have the dog for work. He had looked on the internet and googled various websites and forums where he was advised that, given time, there was no reason why his dog should not turn into a reliable pointer. Sad to relate he was still waiting another year on. Even worse he then got rid of the dog to a bad home because he was annoyed with it. I should have followed my first instinct and homed the dog as a pet where, because it was an endearing enough little soul, it would have been happy and loved.

Interestingly this particular bitch, even from a very early age, was quite unbothered about being lost. She would happily vanish for long periods of time. The insecurity, which should have made her point, was blatantly missing from her make-up.

Nowadays I expect to see some semblance of pointing from an early age. If a dog has not pointed naturally by the time it is six months old, I would start to worry. If you remember, we established that the pointing response is an inhibition of the chasing instinct. The dog badly wants to chase game, but is too anxious to do it. We also said that this anxiety was part of the temperamental make-up of HPRs, setters and pointers.

Therefore, the first indication we may have of potential pointing can be the pup's timidity in approaching strange objects or places. At this stage we should spend time reassuring the pup, until it is confident enough to go forward of its own accord. It will learn that hesitancy is permitted. In contrast, if we can choose a 'gung-ho' approach, and jolly the pup towards whatever has aroused its suspicion, it may learn a disregard for strange things that will lead it to investigate more boldly. Interpret hesitancy as a sign of pointing potential.

You should not rely on instinct alone to develop the point. You can be proactive, both in creating opportunities and exploiting those that do occur. Many HPR trainers will keep **caged game** – a pen of rabbits, a couple of tame pheasants or a cage or two of quail. This exposes the pup from an early age to these fascinating smells.

It will be drawn to them inexorably, and will find it hard to tear itself away from them. Even if you do not have this kind of luxury you can still allow your pup, on the lead, to investigate a pet rabbit that is safe within its hutch. For the sake of the rabbit make sure the dog's first approach is from the back or side of the hutch, so that an over-keen dog does not launch itself at the terrified inhabitant. It is the scent that matters, not the sight of game, so you can even cover a hutch or cage with a tarpaulin.

All you need do is allow the pup to make what will usually be a pretty timid approach. Praise it gently while providing light restraint on a long leash. This is usually no more than the slightest extra pressure but can graduate, depending on the pup's temperament, to quite heavy jerks. However, it should not be so strongly handled that it interprets your control as a correction for finding game.

As the pup comes to, or is brought to a halt, you can reward its caution and hesitancy, using **contact conditioning** to make it stand more solidly (i.e. pushing it gently off-balance *towards* the scent, and from side to side, whilst praising it) – and only checking it if it shows too much inclination to pounce. **Being able to walk up to a dog on point, without it feeling worried by your approach, is a tremendous and essential benefit.**

Once the pup has been made to stand fairly solidly on point, give it lots of vigorous praise before taking it back and repeating the approach. Do this several times over a succession of days. Eventually the pup should not need the restraint to come to a halt, and will also be anticipating your enthusiasm. You should be able to encourage it forward, until it 'sets' itself against your encouragement. A pup that points rabbit in this way will certainly be steady enough when it meets up with feathered game.

Caged game in the field

A lot of HPR trainers will use small cages of quail, partridge, pigeon or pheasant to provide simulated game. Not only does this give the pup the chance to come upon game in a controlled manner but, if two or three cages are placed randomly in cover, it also motivates it to hunt.

Introduce this exercise on a **long line**, which you can hold or allow to trail behind the pup according to your estimation of its waywardness. Because it may be moving more quickly, and may be excited about being loose in the countryside, it will be experiencing a fair adrenalin buzz and may not be as responsive as you hope. The line will enable you to establish your control quickly, in case it is jumping all over some terrified caged birds.

These quail were hand-reared by a dog groomer in her grooming shed, and are totally 'bombproof' with dogs of all shapes and sizes. We have a responsibility of care for these little helpers.

If this is the case, you should correct the dog for being too exuberant, but take it away from the game to do this. Shake it by the scruff, lifting it half off its feet as you do so, and apply plenty of eye contact. As soon as its confidence has subsided a bit you can encourage a second approach on the long line. This time you should be prepared to check any over-excitability, and use quiet praise to urge the dog forward. Keep stroking the dog, and praising it, until it becomes settled. Then move it away and try to get it to find the other birds. This new opportunity is designed to stop it becoming over-obsessed with the first cage. If this happens the dog may start becoming 'sticky' on point, which means it marks game well but is reluctant to flush it on command.

It is worth trying to vary the game species in your cage for this lesson. You want your dog to respond to any gamey scent, and not become a specialist on one species only. An exception to this would be dogs that are intended to point skylarks for falconers to fly merlins at. Skylarks have a strong scent and may spoil the dog for other game. I would recommend anyone considering this chase to get a very steady, more mature dog. A setter that is getting too old for the grouse moor, or perhaps a sensitive Vizsla might suit.

I used both a rabbit and a pheasant pen for training my first HPR, but have never done so since. This is probably because I have been able to take pups out with older dogs that knew the business. The latter will work and ignore the pup but, when they go on point, I can always bring the pup over (having first put it on the lead to prevent a premature flush) and let it discover the scent for itself. This teaches them to investigate cover, rabbit holes, etc.

Working with an adult dog can also teach them the undesirable habit of following the older dog rather than using their own nose. To avoid this, I have found it beneficial to **start pups at a very early age** (commencing as soon as they are clear of their vaccinations), so that they do not have the speed or stamina to keep up. They potter about pretty close to me and educate their own noses. Because they are going more slowly, they very occasionally find something the older dog may have missed.

Dizzying pigeons

An alternative to caged game is the practice of dizzying pigeons. It can be done with other bird species, but the advantage of pigeons is that they go home and can be used again and again.

To do this you hold a pigeon so that its head is tucked under its wing, as it would be when it is asleep. Then circle it round and round in your hands until it is thoroughly dizzy. It can then be placed in cover, where it will remain motionless for a considerable time. This allows you to work your dog up to it, and to achieve a point.

When you are ready you can send your dog in to flush the bird. Your dog's disturbance of its protective cover will awaken the pigeon, which promptly takes wing and sets off for home. It is a good way to teach **steadiness to flush,** as this exercise can be done on the lead or long line. Of course, you do have the added task of maintaining a pigeon loft!

In well-stocked country you do not need a rabbit pen, pigeon loft, or a trained dog, to aid your efforts. Just actively encourage the puppy to investigate every bit of cover instead of strolling about the countryside while it runs wild. Create a team identity. HPRs can be spoiled by an owner ignoring signals from the dog when it has found something. It will eventually despair of you and start flushing and chasing game for its own amusement. Teach it this joint activity is fun, and it will labour away very zealously to find things.

If it sticks its head tentatively down a rabbit hole you can be sure this is occupied. You have time to praise the pup and settle it at the mouth of the hole. Push it forward gently so that it resists. You really want the dog to stand at the mouth of a warren, looking down but not shoving its head down or digging! It is very useful to take your dog ferreting, but you really need the dog to be non-intrusive and quiet.

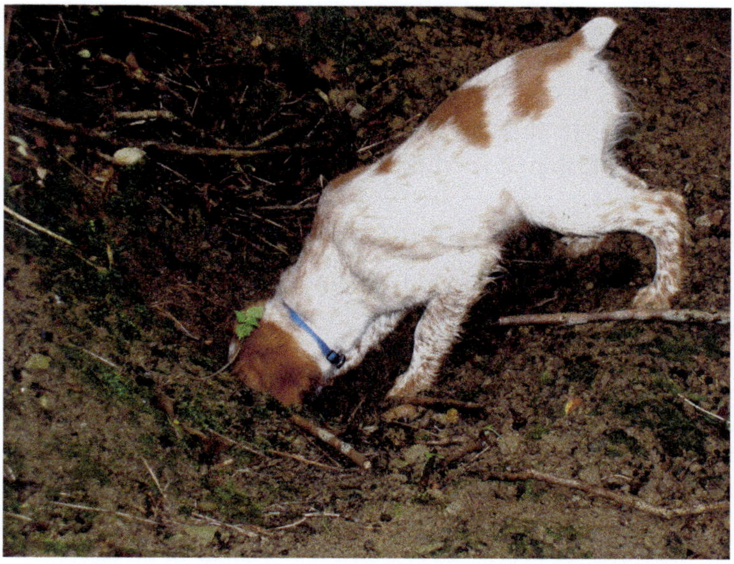

This amount of interest is bit further down the hole than you want. Approach quietly and gently push the pup forward so that it resists the pressure. It will soon start standing back a little!

You may intend to use your dog exclusively on partridge or grouse. If this is going to be your dog's 'specialist subject' you will, of course, dissuade it from showing any interest in rabbits. Here I would definitely recommend you use caged quail as an introduction to pointing game birds. It should then be fairly easy to get a good point on early season birds, as they will still be in a covey. As the pup approaches it will be hesitant, because the scent is coming from a broad area. Being unable to pinpoint a single source it will hesitate earlier and 'stand-off' much better. If it does disturb the birds it will be quite shocked by the sudden explosion and whirring of feathers, and will be more cautious in future. A dog that is worked on game coveys will tend to stand off further than one that is used on single birds and rabbits. Experience will teach you what sort of distance away your dog stops.

An exercise which is done sometimes with very young pups is to allow them to follow the scent of a bird's wing (preferably pheasant, partridge, pigeon or quail), which you trail in front of it with a fishing rod. It will naturally trundle along to investigate this fascinating scent but, as soon as it gets too close, you can whisk the wing up into the air. Because its young eyes are not able to follow such swift movement, it will hesitate more as it approaches the source of the odour. It will give every appearance of pointing but I tend to think this is a visual response that will diminish as its eyesight improves. Try it by all means. If nothing else it will teach the pup to follow its nose and not to race after its quarry as it takes to the wing.

The false point

One of the most irritating vices in a pointing dog is false pointing. This is where the dog goes on point although there is no game present. It is probably indicating where game has been, which is not of particular interest to you. Sensitive and hesitant dogs are prone to false pointing. You can compound this fault by encouraging too much caution while it is still a pup.

I did this with my first HPR, who I felt at the time was quite a forward-going animal. Every time she hesitated, I would urge her to slow down further in an attempt to maximise every opportunity. To oblige me she started to go on point in response to my commands. For the rest of her working life I could never be sure if a point was genuine or not. As she had a nasty habit of working high slopes above me, this entailed very many strenuous climbs up steep hillsides to get above where she stood motionless. As I eventually clambered breathlessly to her level, she would almost shake her head and say '*Sorry, my mistake*' before zooming off to find another scent.

From this I learned that HPRs are not always as out of control as my bitch appeared in her younger days. It is normal for them to run like maniacs and still be working. Given a scent, she would point perfectly well, and it had been my lack of faith in all these things that had made me teach her to false point! Working with her many descendants I have learned that the best approach is to encourage a hesitant dog forward rather than steadying it down. **Urge it into the point and it will stop because it has to. Slow it down and it will point because it thinks it ought to.**

This is where using caged game as a training aid is an advantage, because you know there is something there and can respond accordingly. On open ground you have to trust the dog, which is a difficult step when you are starting out.

It is a good idea to use a light line attached to the pup's collar, but only do this if you have pretty easy cover to work, which affords you good visibility. Alternatively, since your well-prepared pup will allow you to approach it whilst it is on point, you can slip a lead over its head. These measures give you the option of leading the dog forward into the game if it is too sticky, or of checking the headlong pursuit should it be too riotous.

(Photo by Andy Coveney)

At eleven-months this Brittany is happy to hold its point (on caged quail) while I approach. It can easily be put on the lead if a flush is required, so I can then make it sit – encouraging steadiness to flush.

Flushing game

For the novice HPR owner this part of the dog's work is seldom thought of as being difficult to teach. However, in practice, it presents just as many pitfalls as any other aspect. Preparation starts with basic obedience. Whenever you make the dog stay, before free-running it perhaps or before letting it eat from its bowl, release it from the position with the command you will use for flushing. In this way it will associate the command with the freedom to go forward. This ignores the sheer power of the game scent, and some young dogs may be unable to move with such a compelling smell in their nostrils. By contrast, hotter-headed individuals may plunge in immediately on command, and may be tempted to chase the quarry as it makes off. Until your pup has been in the situation, you cannot be sure which way it will react.

In practical terms it often suits you to flush the game yourself, to determine which way it goes. This is certainly true of falconry dogs. The hunting hawk (and most British falconers use hawks as opposed to falcons) does not possess the super-fast speed that the uninitiated expect. In consequence it cannot readily overtake game that goes upwind (your dog will probably be downwind of the quarry and so will flush it that way), nor can it penetrate cover that quickly. So, by selecting which way to send game, you will improve your chances of success.

You will also need your dog to hold its point while you manoeuvre the hawk into the best strategic position. This can take a frustratingly long time. Falconers expect their dogs, in consequence, to hold points for ages, and the longer a dog is expected to hold a point, the harder it will be to get it to flush. For the shooter the problem of speed and direction is less important. Shot travels faster than any game, although a downwind flush will help birds get out of range quicker. Unless you have to wait for a less nimble gun to get in position, the delay between point and flush is much less. Except for the genuine rough-shooter (a fast disappearing breed in the face of the proliferation of driven shooting) the dog handler does not usually shoot, and so is able to concentrate exclusively on his canine helper. This leads to much better control and behaviour.

Ignoring unwanted quarry

This applies primarily to the ubiquitous rabbit that offers temptation to the sporting dog. It can also include hare, muntjac and other deer. Sheep and poultry should not come into this category, and we have already dealt with habituating your dog to them.

Training your pup to ignore rabbits can be achieved by correcting it for showing too much interest in caged rabbits. This can be developed further by using rabbit-fur-covered dummies (or rabbit carcasses) fastened to a long line and hidden in light cover on the training field. These can be dragged out by an assistant while the dog is being taught to quarter. You can also arrange for this distraction to be pulled across the dog's path as it goes out for a retrieve or comes back.

Pulling a dead muntjac about would be a bit more of a challenge and I would recommend that you try to find someone with goats. Their scent is fairly similar to deer, and a tame goat is often prepared to be quite assertive with a young pup. Young foxhounds are sometimes thrown in with a billy (male goat) to learn manners with deer.

Personally, deer have never tempted my dogs so I do not worry about them. They can cause confusion, however. One day my youngster went on one of her very first points at a small bramble thicket. To my chagrin, a pair of muntjac came rocketing out. I was about to tell her off when I realised she was still on point. Closer inspection revealed a fine cock pheasant, clamped down beside an overgrown log. What I had thought was a negative experience turned into a double positive – she had ignored the deer and held steady for the bird.

This lesson was reinforced later on the same day when she did point a muntjac in a bramble patch. On this occasion I was on the opposite side to her, and had chosen to flush the unseen game myself. As luck would have it the deer leapt over her in its efforts to escape me but, getting tangled in the briars, ended up in a heap on top of her. It kicked its way free having coincidentally given her quite a satisfactory battering at the same time. There is no way she will ever be tempted to chase deer after that. Now how can I get a hare to attack her?

Steadiness to shot

In an ideal world every HPR, when it flushes game, will be steady and will not chase. You can teach it to 'drop to shot' where it is expected to drop (or sit) when a shot is fired. This goes back to your early retrieve training, when you taught steadiness by making sure that a retrieve does not automatically follow anything being thrown or launched. This becomes extended by introducing gunshot at the same time as the retrieve dummy is thrown.

I always think that this part of the dog's work is much easier for the shooter than the falconer. The latter has a bird, with which the dog will quickly form a partnership. The hawk learns that it is the dog that is of value to it. It will virtually ignore you while it concentrates on what its four-legged henchman is doing. Equally the dog begins to create a mental pack identity with the hawk. So whenever game is flushed and the hawk (part of its pack) takes off in pursuit, there is a tremendous and natural temptation to join in. This is also the moment when the falconer is definitely concentrating on his bird, and it offers the greatest opportunity for the dog to break its training.

The shooter, in contrast, has no visible chase going on, and no pack behaviour to contend with. The shotgun has a good, loud bang to get the dog's attention. Furthermore, there is no need to run after a hawk to dispossess it of any quarry, or to humanely destroy what it has caught. He or she can stay put, calling the dog back if necessary, and enforcing the stay before sending it to retrieve or, better still, going to collect it him or herself.

Rabbits are the most tempting quarry in this situation, and work done 'running interference' with dummy rabbits is the best basis for obedience. I also use my HPRs for ferreting. When a young pup has pointed a warren, I take the opportunity to tether it to a tree or to my firmly planted stick. When bunny makes a run for it, the pup, if (or more probably when) it tries to chase, will be brought up with a short, sharp jerk. Since I have not been the one to correct it, and it is necessarily still there, I am able to praise it for staying, in the hope of turning the exercise into a positive experience.

The practice of tying your pup to a tree, or tethering it to your stick, has a very valuable spin-off. It teaches the dog to be calm when it is held by the neck. Several times over the years, I have mislaid Brittanys for periods as long as two or three hours, while they were working. Clearly their absence had been spotted early on, but they could not be located. On each occasion I found them in fox snares, which are set at an ideal height to catch the Britts. They had simply sat motionless, just giving the occasional yelp to attract attention. Had they struggled there may have been a less happy outcome.

Being truthful I have to say that this is the weak spot in my training. Partly I forget to tether the pup before it is reliably steady, because I am wrapped up in the hawking and ferreting. Also, I find it useful that my dogs, who I believe are following the hunt rather than actually chasing the rabbit themselves, are quickly on hand if the rabbit has taken temporary cover above ground. They will soon 'bump' it out for the waiting birds. Should the rabbit escape they will mark the hole down which it disappeared, so that I have another warren to work on another day.

Beyond standing up, this young bitch has not tried to join in.

Some of my older dogs also develop the habit of doing a lap of honour round any kill, celebrating it by yodelling with delight. If the kill has taken place out of sight and some distance away, this means they tell me if the chase was successful. It saves me the effort of having to go and look for myself. Incidentally an experienced ferreting dog will point the nearest hole to the rabbit, which also saves time, and will indicate the location of any ferret that is holed up with a difficult rabbit.

Whether you use your dog for finding rabbits depends on your own particular ambition in the sporting field. It is up to you to decide this before you start training so that you know how much to discourage or encourage your dog from the outset.

Hares

A dog which is used for shooting, and is not steady with hares, is going to be a liability. Guns have been known to get careless when excited, and the dog is running in the same dimensional plane as the hare. The frustration of not being able to keep up with its quarry often encourages a dog to 'open up', yelping as it runs. You really do not want this, as all field sports should have a modicum of stealth about them so as not to disturb the game unduly. Worse is the case where some of the leggier, more active HPRs are able to give a fair imitation of a coursing greyhound. How annoying for you to watch clouds of partridge or pheasant rocketing up from a field of kale you intended to work, flushed by your speeding and selectively deaf 'lurcher'.

A dog that succeeds in coursing and catching quarry will seldom be reliable again. It is certainly likely to develop a hard mouth, and can lead it into the unfortunate habit of 'pegging' game. This is different from the keen retriever which will often point a bird which has not risen on being flushed. It may then grab it and bring it back alive. Quite probably such a bird is already carrying shot, and your dog is doing absolutely the right thing.

But what do you do if your dog courses and kills something before bringing it back? You cannot chastise it, as retrieving is a good thing for it to do. Quite frankly I believe you have already lost the battle. Such misbehaviour is rarely an isolated instance and, unless you can quickly re-establish your authority, you could have to consider retiring it to pet status or risk becoming unwelcome on the shoot.

Since writing this a very good, steady bitch of mine tried to retrieve a grey squirrel that the hawks had knocked out of a tree. It bit her quite badly and she has become an inveterate squirrel killer. However, it has not changed her steadiness to rabbit, nor her customary soft mouth when retrieving winged birds. Perhaps I am being unduly pessimistic.

Prevention is better than cure, so do take pains to make your pup steady to rabbit *before* taking it into hare country. You will also find that your dog will often adopt a visibly different position when pointing ground game. Its head will be noticeably looking down, whereas with game birds it will present a flatter, more level outline. If you suspect it is pointing hare, and it is happy to allow you to come up whilst it is on point (as it should be), you can slip a lead over its head.

Multiple points

The very first red-letter day, when my first HPR pointed game for my hawk, fell into this category. I had been working some wonderful parkland which was sown fairly liberally with kale. My bitch went on point under a stag-headed oak, which afforded a perfect perch for the bird. It was the work of a moment to cast (throw) him up on to a convenient branch.

I sent her in to flush and, to my delight, a cock pheasant sprang out of the crop. Quarry and hunter took off and I was unsure what to do. My bitch seemed reluctant to move so I waited to see if the chase was successful. After a short while the hawk returned, having failed to catch the pheasant, and I decided to move on. The dog still would not budge, so I cast the hawk up in the tree again. For a second time I urged my dog in, and was rewarded by a hen bird coming from the same spot.

In my ignorance I had never imagined that this could happen, but since then I have experienced such incidents countless times. This same dog once managed to get five successive points and flushes out of a single covey of partridge. She flushed them so delicately that they mostly flew in ones and twos, while the remainder sat tight. So never assume that one flush is all that there is – watch your dog and trust it. I currently work a bitch who is superb at working the 'flushing points' on our local shoot. When urged in on a few dozen pheasants she will flush them singly, until the keeper sends in his springer to flush them *en masse*. As he says, 'They shoot every bird when they come out one at a time."

Distance of point from quarry

When you first see your dog on point it is a surprise how close to the quarry it has got. Scenting conditions will vary according to wind direction and strength, humidity, temperature and the number of animals or birds found. In all cases they have given off a warm scent that the dog has harvested as it has quartered the ground. As soon as it finds scent it will be drawn towards the source and, as it gets closer, the concentration of moist molecules carrying the scent gets greater. When it is certain it has truly located the source, the dog will, you hope, come to a halt.

Therefore, in good scenting conditions, the dog will be aware of game at a long distance, but will not stop until it is almost on top of it. If scent is poor the dog cannot find it unless it is already fairly close. Instead of working up to a scent, the dog stops dead without any preamble, because the scent was only discernible at the last moment.

A large field that I hunt regularly is entered by a gate in the bottom corner and runs up the hill to my left. Normally we circle this in a clockwise direction, working the hedge bottoms as we go, until we eventually make our way back to the bottom corner opposite the gate. Here there is a wooded gully in which I have a pheasant feeder, and which is also home to rabbits and squirrels.

My bitch knows this exceedingly well, and it is always something of a battle to get her to keep to my plan and ignore what she thinks is the best and most productive bit of land. Very often she will pre-empt any decision by me and head straight for this.

One day we entered through the gate but, instead of careering to the pheasant feeder as she does mostly, or coming in the direction I wanted (which is a less frequent occurrence), she did neither. Instead she headed off across the open field in an unwavering diagonal line that took her to the furthest hedge, some three hundred metres away. Here she went on solid point on a cock pheasant. Its presence had obviously been known to her as soon as she got in the field, if not before that!

Did it jump or was it pushed?

Achieving a good point not only requires a steady dog but also game that is prepared to play its part. Sometimes even a good, experienced dog will 'bump' game up, or you will see birds taking off when the dog has made no effort to go on point. Nervous birds that have been regularly disturbed by beaters and shot at, will often be quick to get up at the approach of anyone. Sometimes the wind direction may have shifted or swirled, or the dog may have been working downwind at that particular moment, so that it stumbled across the game before it hit the scent. Without the facility to freeze the moment and analyse it at leisure, we can only accept it as part and parcel of the mystery of the sport.

Of course, if your dog repeatedly does this you must question its temperament or training. Go back to the caged quarry situation, and exercise stronger control to slow down its approach. Because it is usually the more forceful individuals that do this, you are unlikely to develop false pointing with this corrective training. Review its general behaviour and you will probably realise that its obedience is in need of sharpening up too! Fortunately, we have a long close season in the spring and summer, which is ideal for concentrating on the finer points of the dog's work.

Fault	Cause	Remedy
Dog fails to indicate game	Lack of opportunity Lack of control	Supply caged game or work in well- stocked area Increase obedience sessions Work on long line in controlled Artificial situation
Dog fails to point	Too young Lack of opportunity	Continue to work on stays and to resent opportunities As above
Dog fails to improve over many months despite the above remedies	Lack of inherent temperament	Admit defeat and rehome, retire, replace, etc.
Dog is slow to flush	Oversensitive Overcautious	Boisterous play and treat more robustly to toughen it up Give reassuring praise and delay asking for flush until it is relaxed Put on lead and take it in to flush with lots of praise Practise release from stays with flush command
Dog false points	As above, plus over-controlled	Avoid verbally steadying dog and encourage forward at all times when working Do not praise for pointing (the presence of game is enough reward) but only when flush has been done
Dog chases game	Lack of control	Back to the drawing board on obedience Work on trailing line or use lead when flushing Use of dummy rabbit decoy Practise retrieving yourself – with dog having to wait
Dog 'pegs' game	Lack of control Strong hunting ethic Jealousy	Obedience, staying and welcoming your approach for praise/contact conditioning Use of dummy rabbit Retrieve yourself, plus allowing dog to carry game for long periods instead of taking it off him Practise taking food away

Chapter 8
Distance and Directional Control

Pointing dogs are expected to work a long way from their handler. This may challenge your control over them, so you must start reinforcing their response at increasing distances and build distractions into the simplest lessons. As you get more confident, throw dummies and tennis balls, or even leave bits of food around while doing heelwork or stays. Remember you are in competition with these desirable things so do not be too ambitious too soon. You want to win. Do not restrict the throwing of balls and dummies to long-distance efforts. Articles bowled or kicked swiftly along the ground are much more likely to resemble the movement of rabbits.

Remember that the dog's eventual obedience is driven by its pleasure in the rewards you can give it in the face of all these temptations. These might be food rewards, retrieves, recalls or just very effective praise. Every time you successfully compete with distractions, the good habits you are creating will be reflected in the dog's obedience. This will become evident when it is subsequently working on live game in the shooting field.

It does not take too much imagination to see that creating a sequence of increasing challenges can be done off the lead as well. Strangely you may notice that your dog often behaves better when this is done. Almost certainly this is because your handling will have sharpened up as you lose the feeling of security that a firm grasp on the lead gives you. Learn from this and either apply yourself to being a much better handler in the first instance, or doing everything off the lead. Your progress may be slower but the lessons will be better learnt.

Practising the stays offers another opportunity to reinforce your authority, and develop further the dog's willingness to please you. Expecting a puppy to sit patiently by your side at the school gates, and to allow it to be patted and stroked without moving, may be a bit ambitious at first. Instead of giving it the chance to disobey you it is wiser to avoid exercising strictness at such a time. Go home and get your own children to join in the game, and help you train the family dog. Ask them to make a fuss of it, very gently at first until it learns to ignore them and remains steady and confident beside you. Eventually they should be able to stroke it, and run past with balls or other playthings, without it setting off after them. This sounds dreadfully repressive but, if you have worked effectively on your praise, the puppy will think of it as a game and will enjoy being good. An excellent test of this is if it looks at you for your praise when being distracted. **This achievement is the start of developing really good eye contact at a distance.**

As you may one day expect your trained dog to remain steady to game that it has flushed, it is worth throwing dummies or other articles about while still expecting it to obey. Do not introduce dummies as reinforcement until the puppy has a keen interest in retrieving. Otherwise you may teach it to lose interest in the retrieve entirely. It is always a matter of personal judgement.

This will obviously form part of teaching a more advanced retrieve, but similar interference can be introduced into recall practice as well. Eventually you should be able to throw dummies past a returning dog without it taking any notice of them. You may even get someone to roll one past it from behind, so that it is overtaken by a tempting dummy as it comes to you. When you are shooting or hawking such situations can occur.

Introducing the whistle commands

The human voice has its limitations. As the hunting HPR is normally working upwind, you may find it difficult to project your voice far enough for it to hear you. This will be especially true in windy conditions. A whistle is much more effective, and you should always carry one when you are training, working or exercising your dog. The first whistle command to teach your puppy is the 'Sit'. This is usually given as a single, long blast '*Peeeeep'* – and you would teach the dog to respond to it exactly as it should to your voice. In other words, be prepared to smack it lightly on the rump while giving a quick but gentle tug upwards on the lead. As soon as it obeys you its praise should include the word 'Sit', so that it begins to associate the whistle with the word and the action.

During heelwork use the whistle to stop the dog instead of your voice. Introduce this exercise at other times of heightened excitement. For example, use it to make the dog sit to have its lead on before you go for a walk. Or before you take the lead off at the start of its free exercise. Or when making it wait for its food. The opportunities will be frequent, and you will be creating good habits at times when your control is easily enforced. During heelwork use the whistle to make the dog sit promptly while you continue walking away from it, without a break or hesitation in your stride. In all cases reward the dog ecstatically for its good response.

Stop on recall

The first real test of your dog's obedience to the stop whistle occurs when you try to stop it as it comes to you during a short recall. Until now you have always encouraged it to race up very positively. All of a sudden, the puppy is being made to stop coming to its beloved master or mistress. It will be worried and, consequently, will want to come to you even more in a bid to appease you. When it does sit, rather than asking it to break the stay and come the rest of the distance, it is worth going back to the sitting puppy to make a tremendous fuss of it. Give a little food reward if your puppy is at all interested in food. Then leave the puppy where it is and go back to the original place you started the exercise from. Call it back to this spot and give more praise. Incidentally, the whistle command for 'Come' is generally three quick, sharp blasts on the whistle '*pip, pip, pip*'– so you can start introducing this as well.

With practice you will be able to extend the distance over which you recall the puppy, and also that at which you expect it to stop on the whistle command. For the first two or three weeks keep going back to it rather than getting it to complete the recall after it has initially stopped. As it becomes used to this sequence of events you can throw a food reward to it. This will encourage it to make eye contact with you as well.

Stopping while free-running

Once your puppy is happy with stopping on the whistle, you can introduce this command while it is running free. The idea is that the whistle must be obeyed wherever and whenever it is blown. You are more likely to get a good response if you do this when it is close by you at first. Always remember to go back to it and give lots of praise and/or food. If your dog is really keen on retrieving you can use this as an incentive for it to stop when it is running, by rewarding it with a retrieve. This will encourage it to see stopping as being the precursor of some particularly pleasant activity, and is therefore a desirable thing to do from the point of its own selfish interest.

The stop whistle and the retrieve

This is going to be the first real test of your control. It can only be started once your puppy is pretty obsessed with bringing things back to you. Be careful if your dog is not strongly committed to retrieving – you could put it off even further by introducing this additional control to the exercise too early. With such a dog you will not have much of a control problem anyway, since it is not going to be very distracted with the chasing aspect of the retrieve.

Your work during obedience sessions will have taught the dog to respond to the stop whistle, and you can use this to get the dog sitting before throwing a retrieve article. Since we have already developed the concept of the dog staying on the command 'Sit', it should readily accept that this new signal has the same implications, and will happily wait until sent for the dummy. Don't forget to go and collect the article yourself at times, just as when the exercise was first introduced.

Once the dog is sitting to the whistle before the dummy has been thrown, we can pretend to throw it several times to ensure that the dog is reasonably steady. Then, on the lead if you are unsure about your control, throw the dummy and give the stop whistle simultaneously. If necessary, you can use the lead to reinforce the command. The keen dog will be looking after the dummy, anticipating the excitement of the retrieve. Do not send your dog out for it but, instead, reward it for staying with ample praise and a food reward if that works for your dog. The object of doing this is to get the dog used to making eye contact with you, which means it will have switched its attention to you and away from the dummy. Then leave the dog sitting and collect the dummy yourself.

Repeat this frequently. Ensure that the dog sits every time when the dummy is thrown, and on the sound of the stop whistle. You are trying to make it associate the throwing of the dummy with stopping and sitting. As the dog gets more proficient you can delay the use of the whistle, waiting until the dog has already stopped because the dummy has been thrown. When rewarding the dog with a retrieve, make sure it is settled **and is looking you steadfastly in the eye**, before sending it off.

The ultimate test of your stop whistle (and one I am not sure my own dogs would do reliably) is to send your dog for a retrieve – and then stop it halfway. This can be needed when your dog anticipates collecting somebody else's bird on a shoot day. It will turn to look at you and you must, of course, reward it immediately with your voice. Only once it is settled, and is sitting looking at you (eye contact again), you have the choice of sending it on for the original retrieve (this is a good chance to use the 'Get back' signal), throwing a second dummy for it to get, or simply going and picking up the original bird or dummy yourself. In fact, you may use all three options at different times – just to keep your dog guessing!

For falconers it is worth considering the use of a pair of hawk bells as a signal to stop, in exactly the same way as I have described for training to the whistle. Simply rattle these whenever you want the dog to stop. This means that they should become steady to the hawk or falcon taking off after game the dog has flushed. This conditioning can start even when you are manning your bird. Should it bate from the fist there will be a reasonable amount of noise from the bells, and you will automatically stop to allow the bird to right itself. Take the opportunity to insist on the dog stopping and sitting when this happens

Stopping at a distance

In training any animal your job is to create a suitable situation, in a favourable environment, to get the best result. Blowing the stop whistle, as soon as you first let the dog off the lead in its habitual free-running area, is doomed to failure. Instead choose ground that is unlikely to hold game, and wait until the dog has run off a bit of steam. Now is the time to use your single long blast on the whistle, while the dog is no more than twenty metres away. If you have done your groundwork properly it should respond, and be looking to you for praise. Equally, because it is further away, it might creep towards you a bit. This is not bad as it is focusing its attention on you, and will certainly be making the required eye contact. At some distance from it you should verbally praise it really well, *before* walking back to the dog and making a huge fuss of it (with food again if you think this is appreciated).

Before releasing the dog, to run off and play again, it is worth giving the 'Get back' signal and command, and make it move back some five or ten metres. Use the stop whistle again, and give even more praise before releasing it from your control. If you have generated sufficient enthusiasm with your voice, the dog will look forward to this new game and will sit quickly in anticipation of your reward. With practice you will be able to extend the distance (and thereby the range of your control). It will get into the habit of enjoying itself, but still keeping one ear open for your whistle. One of my dogs really likes being redirected – giving an excited yelp whenever it set off in its new direction.

Failure to respond to the stop whistle

There are some basic reasons why your dog may not respond promptly to the stop whistle at a distance. It may simply have insufficient respect for you, and is relishing being beyond the perceived area of your authority. This will certainly be true if it ignores you completely. It is telling you that both your praise and your correction are inadequate. Recognise that you have been too ambitious, and go back a step or two in your training, concentrating on those situations where you can control both distractions and temptations.

You can use an assistant to counteract this. Start with them having the dog on the lead, but not in any way being expected to exercise control over it. They can just be standing around with it while you are doing other things some way away. When the dog is not particularly paying attention to you, give the long single-whistle signal to 'Sit'. Your assistant can quickly give the upward jerk on the lead, and the swift smack on the backside, to ensure prompt obedience to your signal. This should be done without any verbal command from them, so that the dog learns to associate their control as being triggered by your whistle.

This understanding is reinforced if you give the deserved praise for their obedience, instead of this coming from your assistant. Repetition should bring about further improvement. Eventually you can dispense with the lead and rely on the assistant's presence (and possibly their use of voice) to support your whistle.

Coming towards you on the stop whistle

An irritating habit is where the dog stops running in response to your signal and whistle but, instead of sitting, creeps slowly towards you. Previously it has always started retrieves and heelwork from a position by your side and is being obedient in its own way. It is anticipating doing something exciting with you. Confusion can also arise because you may have sometimes used the stop whistle unsuccessfully, and have then resorted to bellowing at it by name, which you will often use when recalling it. Try to be very consistent and sparing with your use of the whistle to avoid such misunderstandings. **Do not become 'whistle-happy'.**

This problem can be remedied by using both seen and unseen (or 'blind') retrieves. Once the dog has finally consented to stop (and has been adequately praised) you can throw a dummy either behind it or at a right angle to one side. **This will remove any advantage it may have gained by coming closer**. When throwing to the side, always try to land the dummy so that it is behind a line level with where the dog is sitting. If it is anticipating having to retrace its steps for its much-desired dummy, it will soon look forward to this and will not want to come towards you. Alternatively, you can plan to stop the dog by a gateway, a convenient gap in the hedge, or a 'tramway' in a standing crop, through which you have previously placed a blind retrieve. By doing this you hope the dog will realise that a prompt stop will speed up this desired, but unexpected reward. If the dog comes to believe that obeying you turns out to be in its own interest, you can be sure it will be better motivated than any wish to please you will ever achieve.

As you get more confident of your control over the dog you can throw dummies while it is running, and give the stop whistle as it sees the article flying through the air. Successful development of this practice will lead to the dog seeing the movement of the dummy as a situation in which it is expected to stop. When this understanding transfers to the shooting or hawking field you are well on your way to having a dog that is 'steady to flush'.

Directional commands, whistles and hand signals

There will be many occasions when you will want to direct or redirect your dog. In each situation the command, whistle and hand signal must be consistent if you are not to cause confusion in your puppy or dog's mind. At the risk of stating the obvious they are as follows:

- **Towards you** The command is 'Come' and the whistle signal is three short, sharp peeps '*pip, pip, pip*'. Different handlers use different hand signals. Personally, I simply lower my right hand in front of my right thigh as an invitation to come. This is because I will often have a hawk on the other fist, which makes me effectively one-handed. Other people will pat their legs with both hands in a similar invitation, and will crouch down to make their stance less forbidding. Yet others will stand tall and stretch both arms out wide above their heads, to offer a very clear signal for dogs working a long way away, or in high cover that would obscure the visual signal otherwise. Whatever you choose, try to introduce it consistently and keep to it.

- **Stop or Sit** The command is 'Sit' or the long single blast on the whistle *'peeeep'*. Inevitably the dog will turn to face you and the hand signal is an upraised hand, palm towards the dog.
- **Directly away from you** Nobody seems to use a whistle for this. The verbal command is 'Get back' or 'Back' – **and it makes sense to use this even when you are sending the dog away from your side.** The dog will interpret this as going directly away from you. For this reason, I think it is useful to have the dog facing you at the start, while you stand in front of it – and directly opposite to the direction you want the dog to go. The hand signal is a pushing motion towards the dog with the open palm of the upraised hand. Rough-shooters may choose to use their left hand for this as they will be carrying a gun in the right while, conversely, falconers will use their right because the hawk is sitting on the other.

The back signal

- **Left** This too has no verbal or whistle command. The hand signal is initially an exaggerated gesture with the outstretched left hand and possibly a body sway in that direction as well. Falconers may use the right hand across the body in a similar manner, as the hawk will not take kindly to being waved about as some kind of feathered semaphore flag. For extra effect a beating stick will increase the visual size of this signal. (NB If you wish to trial your dog, you should not rely on your stick, as you will not be allowed to use one in competition.)

As the dog becomes familiar with the situation the need for exaggerated signals is much reduced, and the merest nod of your head or flick of your hand will be sufficient. Unconsciously you will almost certainly rely increasingly on your body position, and will influence the dog just by turning very slightly in the direction you want.

- **Right** The opposite to the above, with the exception for rough-shooters that they may choose to use their left hand across the body, or gesture with the gun itself towards the right. Whichever they decide on must be used consistently from the outset.

- **Turn** There is no vocal command for this, and you use only a single, sharp *'pip'* on the whistle to attract the dog. This signal is given just as it reaches the end of its outrun to either side, and should bring it back towards you. In the early stages you can give the left or right-hand signals, as described above, to emphasise the required direction. You will quickly start to rely on your body position. The turn signal is really a watered-down version of 'Come'. The dog runs across you rather than coming, because you have not used the insistent triple *'peep-peep-peep'* and because it wants to continue hunting.

Often a hard-running dog will turn at less than a 90-degree angle, instead of coming back squarely across you. Just let it keep running until it is almost crossing in front of you – and then repeat the single *'pip'* to turn. At the same time turn your body even further in the required direction and check your pace. This will encourage it to take an extra 'bite' and come back to the line you want.

Because nobody wants to disturb game prematurely you need to be stealthy. It is advisable to use your voice and whistle as quietly and as little as you can. The more obedient your dog is, the less you will need to direct it. As it learns the job it will begin to react automatically to the situation as a result of its past experiences. Learn to be patient and let it work things out for itself. If you have trained it well it will be trying to please you, and will not need continual urging on.

Directing your dog to a retrieve

Dogs have a remarkable ability to remember where shot birds or dummies have landed, and you will have very little difficulty encouraging it to sniff out something it has seen fall. When it does this it is said to have 'marked' game. However, you can often be asked to retrieve fallen birds from an area which your dog has not previously seen. These 'blind' retrieves will depend very much on two things. Your ability to direct the dog to the area described to you, and the dog's determination to find anything for you. It is important that you can direct your dog to a point **where it will be downwind of the fallen game**.

Getting distance

Developing your dog's willingness to run away from you in a straight line requires you to give it increasingly distant retrieves as it gains confidence.

Start with 'memory retrieves', which consist of the dog returning to a known spot which it associates with dummies. You can do this by getting an assistant to throw a dummy for your dog at a distance, after first making a noise or perhaps firing a starting pistol to get its attention. Line the dog and yourself towards the dummy and send it with a forward hand signal. While it is returning to you, your assistant can now drop another, unseen dummy in the same spot. The dog will not know this but, if you line it up and direct in a similar fashion, will probably go back to where it previously found its dummy. Sometimes your assistant may fire the gun as a precursor to this so that it is anticipating finding something in this area.

As the dog associates being lined up and directed with finding its favourite article, it will soon expect to find it over an ever-increasing distance. If you do not have an assistant you can simply drop a dummy quietly behind you as you walk along a path. Send the dog back for it, initially while the dummy is visible but developing the distance as it becomes more confident. It will not run in a ruler-straight line and will have no idea of how far it has to go. Any hesitation should see you immediately repeating the command and the 'Back' signal, until it has got to an area downwind of the dummy. Then repeat the command, or use whatever signal or whistle you intend it to associate with looking for a retrieve in that vicinity.

To teach a more controlled response depends initially on your ability to stop it promptly. Having achieved this, you need to be able to send it purposefully where you think it should go. A good way to train this is to use a closely grouped cluster of dummies, perhaps five or six together. Leave these about fifty metres away but in view. Tell your dog to sit, and then retreat a couple of metres only, before facing the dog. Give the 'Back' signal towards the pile of dummies and tell it to fetch. While it is running out you can retreat a few paces further back so that each retrieve is to the self-same place, but returning over a slightly longer distance. With practice you can develop this exercise until the dog will happily run out for more than one hundred metres.

I find that, increasingly, I dispense with any command at all for 'Fetch' – as the dog needs only to be given the required directional signals. Since we only use these to help it find something, the use of them will tell it that we expect it to do just that. When the dog is 'getting warm' I then introduce a short undulating whistle *'pee-weep'* which, from experience, the dog associates with searching the immediate area. This is introduced from the earliest retrieves as the dog gets really close to the article. If it moves on too far in any direction, I can still stop it, redirect it and repeat the *'pee-weep'* whistle command again.

Some handlers seem to enjoy complicating things and use a variety of different commands for different aspects of the retrieve. To them, quite clearly, 'Fetch' only means the dog is to go and retrieve a marked bird or dummy. With a blind retrieve these handlers often introduce a new (and superfluous) command 'Hilost' to indicate the need to search for something unseen. To indicate that the dog is 'getting warm', the handlers stop their dogs and then use yet another command 'There' to indicate that they have arrived at the correct place, followed again by 'Hilost'. NB Different handlers will use different permutations of these and other commands.

To send a dog in any direction you must be able to make it turn at right angles from a stationary position. Difficulties arise when the dog thinks it knows better than you. It will start running in ever-increasing circles and zigzagging about the ground to pick up the scent. Should it do this for long enough there is a good chance that it will eventually find the desired object, and it may become convinced that this is the correct strategy in this situation.

You can also leave the dog facing in the direction the dummy has been placed or thrown. Move away from it to your left. You will now be at right angles to the line it needs to take for the retrieve. Here the 'Go left' hand signal can be used, and the dog will learn to move off at right angles to you. Repetition will allow you to be both further from the dog in giving the signal, and further from the dummy.

After a few successful retrieves you can change to the right-hand side and introduce the 'Go right' signal. Once again practice will lead to a positive and confident response from the dog.

A progression of this exercise is to teach the response to the 'Go left' and 'Go right' signals at a distance. Use a hedge, wall or fence and place a hidden dummy or two at the base. Sit the dog about thirty metres out into the field and direct it straight to the hedge, where it should find the dummies. It will soon go confidently towards the hedge in anticipation of its reward. On subsequent days vary the destinations to other places in the same field, so that it is your signals, rather than any specific spot, which it is responding to. Now place two piles of dummies, unseen, to both the left and right of the central line, perhaps twenty or thirty metres apart. Send the dog as before, but stop it as it hesitates on reaching the hedge. Then direct it towards the hidden dummies. Vary which way you send it, so that it cannot guess and must wait for your directional signals. With greater experience you can increase the distance. Eventually you should be able to send the dog to a far distant point, stopping it on the whistle before it gets there, and then redirecting it to dummies out in the open field.

Be prepared for the dog starting to see hedges, fences or walls as signals to turn, and counteract this by sometimes sending it through or over them for retrieves.

Runners

If your fallen bird is a runner your dog will have to track the route it has taken. You will not know exactly where the bird has gone, nor how far, so your ability to help is very limited. My feeling is that it is better to keep quiet and allow the dog to use its nose and its initiative. Only if it is really not getting anywhere should you think of stopping it and redirecting it to the place where the bird was seen to drop. Otherwise you run the risk of creating an automaton that is devoid of initiative and totally reliant on your guidance.

You can use a freshly shot bird's carcass to create an artificial 'drag', rather as foxhunters are now required to lay trails for their hounds. However, this raises the possibility of your dog following the scent of departed game, when it should instead be quartering for quarry that is clamped down in cover. Without knowing what the dog can smell you have no idea if it is working correctly or if you should make it stick to its pattern.

Chapter 9
Competing

When we first select our puppy, we initially choose it for the achievement and excellence of its parents. To its owners, each pup should be 'the best dog in the world', but the truth is that they cannot all be world-beaters. Some dogs are so indifferent in the quality of their work, appearance, temperament and vigour that we would happily pay to avoid any close relative of theirs. Yet their owners may still praise them to the sky, and may subsequently choose to breed from them. Of course, this is their own business, and they may see virtues in their dog we have failed to recognise. However only the most informed judgements will help us reach a truly objective opinion, and this can only be obtained by comparison of like with like.

If we consider the many packs of foxhounds that exist in Britain, we can see this selection at its most ruthless and most effective. They weed out those hounds which fail to reach their exacting standards of performance and which will therefore be culled. This is possible because they are in a position to compare their young entry on a day-to-day basis in identical conditions. Equally they can recognise those individuals which consistently outperform their peers, and which they will retain as part of their breeding programme. What helps even further is that there are no pet owners involved in this decision-making, so that sentiment has no influence.

In many ways all good, professional breeders follow these principles, and only breed from the best, never using sub-standard stock. Comparing retrievers and spaniels is relatively easy on a driven shoot because there will be several Labradors and springers etc. working each day. Any keen shot, going from shoot to shoot throughout a season, will see a lot of the same dogs performing week after week. Enquiries will soon help identify related stock that work well for a variety of handlers. Equally it will also highlight those handlers who always have good dogs, when other related animals fail to impress in less skilled hands. This knowledge is just as important, as you cannot buy the trainer, and will be unlikely to equal his or her skills.

By and large, pointing dogs are not normally seen working in any numbers at the same venue. Their style of work means that they are solitary hunters rather than operating in packs. Therefore, the prospective HPR owner, and the would-be breeder of HPRs, need to find ways of sourcing good stock. Happily, there are events which you can attend, that give you this opportunity.

Most of the gun-dog training clubs, and all the various HPR breed clubs, stage different levels of working competitions and activities. This encourages the retention of the breed's working characteristics, and also provides a platform for the more competitive owners to enhance the reputation of their kennel. As a handler it can provide a valuable 'second opinion' on the effectiveness both of your training programme and of your dog's performance.

TAN (*Teste d'Aptitudes Naturelles*)

This is the simplest measure of a dog's working ability and requires no skill from the handler at all. It was developed by the French breeders of the Brittany to help sift 'the wheat from the chaff', and other HPR breed clubs have adopted this. The format is uncomplicated, and the event is held on a single large, open field. Usually this will have a good covering of rough grass or growing crops, but very little in the way of challenging cover. To ensure every dog has the chance to find game, the organisers will discreetly hide two or three cages of quail in different parts of the ground. Entry is customarily, but not necessarily, restricted to pups and young dogs between the ages of six months and three years of age.

Each dog in turn is called forward to the judge, who will advise novice handlers which direction (according to the wind) will be the best for the dogs. Judge and owner will then stroll gently over the field, allowing the dog to do what it wants. It is not a test of handling. What the judge is looking for is those dogs that have a hunting ethic, and can independently show an intuitive use of the wind.

Inevitably the dog will at some stage arrive in an area downwind of a cage of quail. Its reaction is important. Firstly, it must respond to the scent that it has found and then, ideally, point the quail solidly. Lacking any experience, many youngsters will have difficulty with one or both of these achievements. The judge will offer constructive help to the handler, and may even treat it as a training exercise to see if the dog can be 'switched on' to its job. Many youngsters demonstrate a ready enthusiasm for their role and this, in turn, encourages their owners to feel more confident of trying to train their dog for work. A dog can fail its TAN for showing no work ethic, a lack of response to the scent of game and a lack of steadiness when finding game. Unruly dogs, which demonstrate no respect for their owner and are deemed to be out of control, will also be discarded. The judge will select winners from those which have hunted with zeal, instinctively used the wind, remained responsive and pointed their game. His criterion will simply be 'Has this youngster got the basic instincts that lead me to think it would be a good working dog?'

Judge and TAN competitor setting off with a young dog

The TAN is a good social event, as everybody likes watching young stock coming out for the first time. Breeders are reassured if their produce performs well in the hands of inexperienced owners, as this validates their breeding. This competition does not require any skills from the dog that might have to be taught, e.g. retrieving and flushing. There is no limit on the numbers that can be run in a TAN.

Spring Pointing Tests

This is a more advanced exercise in which dogs are asked to work ground to find, point and flush partridge and, less commonly, grouse. The number of competitors (or runners as they are known) is limited by the amount of land and quarry that is available. As the name suggests these tests take place after the shooting season, and before birds have started nesting, i.e. in late March to mid-April. Here the dog is expected to quarter a designated area, and to find and point game. The judge will be looking for style, thoroughness, outrun and steadiness to other game (rabbit, hare and deer). When the dog finds birds, it must hold its point and allow the handler and judge to approach. On the handler's command it is then expected to flush the birds, but not to run after them. As the dog will also point other game it may be asked to flush these, but remain steady. Steadiness means that the dog will not pursue a hare for more than fifty metres! No game is shot so pointing tests do not include retrieving.

These tests are run by the various HPR breed clubs, and many people join several clubs to increase their chance of getting one of a limited number of places. Priority is normally given to Club members and dogs of that Club's specialist breed. There will be a reserve list of competitors drawn up, who may get a run at the end of the day if there is sufficient game to provide the opportunity.

As with any activity based on hunting you must be prepared for the extremes of weather that Britain can throw at you. There may be no shelter, no toilets and no catering so you will need to consider these problems as well. Because the event is held on private ground by the owner's invitation, it is considered rude for guests (you in this case) to leave early.

Grouse Pointing Trials and Tests

Pointing trials are run in late summer (just before the grouse-shooting season starts on August 12th). They enable dogs to be assessed on hunting, pointing and steadiness. Mostly they are staged by setter and pointer clubs but some HPR clubs run them as pointing tests. No game is shot so retrieving is not required.

As with the spring pointing tests it is not always easy to get a place and, if your dog is not pretty competent, it is considered bad manners to take up a place that a more experienced dog could have had. There are few spectators, and the 'gallery' will mostly consist of other competitors waiting for their turn to run their dogs. To avoid them and their dogs interfering with the dog being judged, the 'gallery' is kept so far back it can often be difficult to see what is going on.

Grouse counting

This is not a competitive activity, but is run to help keepers establish the population and whereabouts of the grouse on the moor. Counts can be held in early spring, after the shooting season and before the birds get down to breeding (through February and March). This identifies the number of adult birds that have survived the winter and will, potentially, be breeding in the summer.

More usually, counting takes place in the late summer (latter half of July and first week in August) to help predict the numbers and distribution of the population and therefore the potential 'bag'. The keepers will also learn the average number of young birds in a covey, which will not only reflect the weather during the breeding season but may also indicate the local parasite load or predation level.

Working dogs on grouse in this way, particularly youngsters, is an excellent education. It can be hot work and many moors are remarkably short of running water for your dog, so take some for him as well as yourself. Another precaution is to know the nearest vet's surgery as adder bites are far from uncommon. Coping with deep heather can be exhausting for the dog (and you) so it is sensible to rest your dog periodically while another is working. Heather stems are quite abrasive to the dog's pads, particularly if the presence of moisture leads to a 'wet and dry effect'. Do keep an eye on your dog for thin and sore pads.

The technique in grouse counting is to divide the area into workable beats, which have identifiable boundaries, after the keeper has told you which bit of moor he needs counting. He will know which beats are typical of his ground and, by recording their population of birds, can estimate the productivity of the moor as a whole. Always start from the downwind boundary if you can. Leave a marker at the point where you are going to start from and also make a note of the distant landmark that you are going to work towards (a brief sketch of the skyline seen from the starting point will suffice). Take your time and methodically work as broad a front as you think the dogs will maintain. Once you have covered your ground, return to the start to work into the wind again on the adjoining beat. Note down the numbers of birds and their exact location, and also if they go in any favoured direction. This information should be passed on to the keeper by telephone or by calling round in person.

You should do some grouse counting before trying to run your dog in a grouse test. Counting in early spring is less productive as there will only be singletons and pairs on the moor. These are very flighty and you will be unlikely to obtain many solid points. Never mind, it will switch your young dog on to the scent for later in the year.

In the summer the scent from a large covey is so widespread that your dog should stop well short of the general vicinity of the covey. Since you have to flush the grouse to count them, this offers a great opportunity to work on your dog's steadiness. You can choose to perform the flush yourself at times, so that the dog is kept guessing and does not rush in when you approach near its point. Conversely, with a dog that is sticky on point, you have the chance to let it run in quickly, because you do not have to wait for a gun to get in position. A careful dog should manage more than one flush per covey, putting up just a few birds each time it 'makes in'.

Above all your mistakes can be made in relative privacy, and nobody is upset at losing the chance of a shot.

'Dogging in'

As with grouse counting this is not a competitive activity, but it is incredibly valuable for gamekeepers. More or less as soon as pheasant poults have been released from the rearing pens to wander about the shoot, they show a remarkable ambition to disappear over the shoot boundaries on to your neighbour's land. Where this is also stocked and shot over there is not much of a problem, as the two populations simply balance out each other's gains and losses. However, it is annoying when a substantial percentage of your young bird population migrates to unstocked land.

Anyone with ambitions to work their young dog on their local shoot should approach the keeper, and ask if he wants any 'dogging in' done. Offer to patrol the boundaries, at whatever time he thinks is most rewarding, so that you can turn back escapees to the safety of their own home. HPRs are brilliant at this as they will point the young birds, giving you the chance to flush them in the correct direction. It also gives you the chance to practise the delicate matter of your dog's steadiness to flush – at a time when nobody is about to watch any disasters. Because of the sheer numbers of birds that are put down each year, you are very likely to be able to repeat this exercise many times during a short walk.

As you and the birds get more practised, the latter will often anticipate being flushed and start wandering back in increasingly large groups – starting with twos and threes but quickly gathering in numbers until you may have fifty or so being herded up a track in front of you. Your dog should 'stalk' these like a collie, keeping them on the move but not actually pushing them along. This is good preparation for working with red-legged partridge, which have an annoying habit of just edging ahead of your dog through standing crops. You will know exactly what the situation is by watching your dog's careful and hesitant progress, as it keeps in contact with the birds but does not put them up.

'Dogging in' can run from early August to the end of October – or even later if the shoot boundaries are particularly 'porous'. The keeper will tell you when the need is over. Don't expect to be welcome if you only do the occasional hour or two to suit yourself. The keeper needs to know who is on the land and when, so always communicate fully. Also tell him or her of predators you may see, or just simple things like feeders needing topping up.

Working Tests

These 'out of season' events simulate, as closely as possible, a day's shooting but without game being killed. Instead the dog will be expected to work ground as if for live game, as it did in the pointing tests. It will also be asked to retrieve in a variety of situations. Depending on the level of the competition this will be for dummies and/or cold game (i.e. thawed carcasses from the freezer). These can be as 'marked' retrieves, where the dog can see the retrieve article drop, or as 'blind' retrieves, where the dog is sent into cover to locate a dummy or a hidden carcass it has not seen drop there. In the lower levels of competition, 'marked' retrieves can be from water, while in higher stakes the dog will be asked to enter water and cross it to seek a 'blind' retrieve.

As a general rule the more advanced the test the greater is the examination of the handler's control of the dog. This is especially demonstrated in his or her ability to direct the dog towards a 'blind' retrieve, irrespective of the distance required, or to perform 'split' retrieves where the dog has to collect the nominated one of a number of dummies or carcasses. Although no game is shot it is normal for blanks to be fired in conjunction with the use of a dummy launcher. This not only tests the dog's reaction to gunfire, but also sets up the situation where a dog must remain steady to shot.

Field Trials

These are as close to the real thing as you can get, and are held in the shooting season. Instead of dummies and cold game everything revolves round actual work. Your dog will be expected to hunt cover thoroughly and stylishly, and should not miss anything, i.e. 'blink' game, or false point. It is the handler's responsibility to tell the judge if he or she is happy that the dog is on point. A gun or guns are then positioned and, when the judge is ready, the dog should flush the game. It must 'drop to shot' or at least remain still until told to retrieve the game.

It is quite common for a dog to be eliminated after just a few minutes for missing birds or failing to point them properly. (Sadly, the dog might subsequently have steadied down and worked well, but it will not be given the opportunity.) Others may fail to retrieve, or may mouth their game, or misbehave in other ways. After all the competitors have had one run, those that have not been eliminated will have a second go. Often the winner is the dog that has had a lucky draw, getting good ground and steady birds. Another, potentially better, dog may be faced with difficult cover and unsteady birds or runners.

To win a field trial takes not only a good slice of luck, but also a dog that is well-trained enough to take advantage of that good fortune. You need to be prepared for many disappointments before the capricious hand of fate tips the scales in your favour.

HPR competitors and their dogs wait to complete the water retrieve at this open trial

Apart from the winner and the placed dogs, the judges will also award Certificates of Merit where a dog has shown all the required skills, but has simply been beaten by better (or luckier) dogs. In the dual-purpose world of HPRs these certificates enable show-winning dogs to upgrade a Show Champion title to a full Champion, which is, quite rightly, highly regarded. If you look at your dog's pedigree you will see that some are 'Ft Ch' (Field Trials Champions), others are 'Ch' (see above), while some have 'Sh Ch' (meaning their success has been restricted to the show ring).

Field trials, and the other tests described above, are very sociable occasions. Because all the competitors are aware of the element of luck that exists, they are normally supportive of each other. They have also learned to be philosophical in defeat, as just one unfortunate error at the wrong moment will lead to even the best of dogs being eliminated. Attaining the required perfection is so hard that it is common for no first prize to be awarded at all and, at the first trial I competed in, every dog was eliminated! Hats off to those diehard enthusiasts who keep coming up for more.

There are different grades of Working Tests and Field Trial. Tests commence with Puppy Stakes (i.e. for dogs of six to eighteen months of age). Puppy Field Trials are never staged as it is thought unlikely that there would be sufficient well-behaved and experienced youngsters to justify this.

Graduating from this are the Novice Trials and Tests. Judges will be more sympathetic at this level and will allow much more leeway where small faults occur. Moving up a level we come to the All Age Trials and Tests where any dog may compete. These are generally accepted as being a halfway stage towards Open Stakes. Judges will expect and demand a higher standard of work from both you and your dog.

Finally, there are Open Trials and Tests (these, although theoretically open to anyone, will normally restrict entries to dogs that have won at Novice or All Age level). Expect to see higher standards and tougher conditions. The water retrieve is a 'blind' retrieve across water. Championship Trials for HPRs are rarely run through lack of suitably qualified dogs. To become a Field Trial Champion a dog must win two Open Trials. Before entering a trial, do go as a spectator to find out about the way they are run.

Kennel Club 'J' Regulations

These are the published rules relating to Working Tests and Field Trials. They are under continual review, and cannot therefore be given here with any confidence as to their accuracy. Anybody who fancies running an HPR needs to become familiar with them, or have a friendly expert to guide you through the labyrinth.

Conclusion

Not every HPR owner will want to test his or her dog's ability (and their own) in the competitive arena. However, taking part at some level will ensure that you have the means to make a comparison between your dog's ability and that of others of the same breed. Without this there is always a danger that your breed will slowly slide into a division between show/pets and working dogs. Even attending as a spectator is worth doing, as it will extend your knowledge.

Chapter 10
The Theory of Dog Breeding

I was a child when we still had Religious Education, not just in morning assembly but as part of the weekly timetable of lessons. A familiarity with the Holy Bible (King James version, of course) helped us find those books where the slightly naughty word 'begat' was repeated endlessly. Even in those far-off Old Testament days it was important to remember the lineage of kings and lesser mortals. At grammar school I learned about Charles Darwin, his theory of evolution and natural selection, and also the discoveries of genetic inheritance first shaped by Gregor Mendel, the Augustinian monk.

It was the work of these two pioneers which really triggered the beginning of dog breeding as we know it. Dog owners, particularly wealthy sportsmen, realised for the first time that they could establish identifiable breeds to suit their own personal needs and preferences. These breeds were usually an attempt to fix a type that had been developed to work in the local countryside, and were ideal for the quarry and style of hunting that particularly appealed to their patrons. Often an individual dog, held in high regard, was used as a 'template' for their efforts and became the basis for the breed standard. Initial selection of stock was by physical and functional resemblance, and this is known as choosing **'photo-typical'** parents, i.e. parents that looked and performed the same. By a process of selection (and pretty ferocious culling) their resulting progeny began to breed true to type and to achieve the desired uniformity, so that they became predictable in looks and performance. The behavioural and physical traits were now fixed in the genetic blueprint of the breed, and such dogs are said to be **'geno-typical'**.

The influence of inherited behaviour is increasingly recognised. The importance of selecting breeding stock for temperamental traits is as important as seeking physical excellence – and avoiding physical exaggeration!

Pedigree

Recording the breeding and bloodlines was sometimes very casually done, but slowly a term began to be used that defined those dogs of known ancestry. They were said to be **'pedigree'**, which still only means that its ancestors are known for the last five generations. Today, in some breeds that have a very small genetic pool, it is sometimes considered acceptable to introduce different qualities, or remedy common faults by crossing with a dog of a separate but similar breed. As long as their breeding is known, the resulting progeny can be legally called pedigree, even if they are cross-bred.

Because a dog has a pedigree there is no guarantee that it will be anything more than a general, and sometimes quite vague, approximation of what the breed standard has led you to expect. A pedigree is merely a record of how your dog is bred, and it can just as easily be a record of bad breeding as an indicator of potential excellence. Moreover, a pedigree is as useful as the honesty of the people who filled it in! It only takes one rogue to invalidate the entire document, and there are rogues in the dog world as in any other walk of life. It is one of the most important roles of the Kennel Club (and its overseas counterparts) to maintain accurate records of breeding but they cannot know all the truth. Therefore, even a KC-registered puppy may have a false pedigree.

Example 1 A person wants to mate their much-prized bitch with a well-known stud dog, because of the many good qualities he is known to pass on to his progeny. Unfortunately, the dog is now getting on in years, and has intermittent arthritic problems that make it difficult or painful to mount a bitch. The day the bitch arrives is one of his bad days, so the owner of the bitch is told that the dog does not like 'performing' with strangers around, and that they will have to take the bitch to the mating shed, where he will be more likely to carry out his duties. Once there they substitute a son of the intended stud, who they are anxious to have used. Of course, the mating form is filled in with the old dog's name on it, and the pups will be registered as his sons and daughters. If any of those dogs are subsequently bred from, the whole pretence is perpetuated. This happened to a friend of mine who was breeding for a specific, predictable colour, and the fraud came to light when the pups arrived and were the wrong colour. This was before DNA testing was available but the resulting court case established that the swap must have taken place.

Example 2 A keen shot has acquired a dog from a friend but, unfortunately, the papers have been mislaid. The dog turns out to be a superlative worker, and equally good-looking. A lot of people are anxious to use him at stud and the owner is presented with the opportunity of making some money, and also of seeing his excellent dog influence the breed. Fortunately(?), he knows someone who has a castrated pet dog of the same breed, and gets the pedigree and other papers from this owner, so that people using his dog can register the pups. Even if its progeny are truly admirable, and a credit to the breed, it still means that the pedigree is worthless. One danger, of course, is that the dog may carry a potentially lethal, but recessive and therefore hidden genetic health problem. If, unknowingly, a very closely related bitch is brought for mating, there is a high risk that any offspring will be affected.

Knowledge of pedigrees can only be acquired over a long familiarity with a given breed. Anyone with an ambition to breed dogs should try to find a 'guru' who can tell them what names add or subtract value from a pedigree, and why. Never stop studying pedigrees and watching as many dogs of your breed as you can. One day you can be someone else's 'guru'.

Cross-breeding

Cross-breeding occurs where dogs of different breeds are mated together. This may be done deliberately, e.g. in the case of Wire-haired German Pointers being introduced into Hungarian Vizslas to create a more hardy, forceful dog, or where Labradors and poodles have been crossed to produce the Labradoodle, a dog with a great pet temperament as well as an anti-allergenic coat. You should never contemplate embarking on a programme of cross-breeding, unless your breed club has sanctioned this and has notified the Kennel Club of your intention. It would take an extreme problem to make this desirable. At one time the Clumber Spaniel had an extremely narrow gene pool, and a host of inherited health issues. It was agreed that English Springer Spaniels could be introduced to solve both these problems. The result is the modern 'working' Clumber, which is much more active and robust than its ancestors.

Cross-breeding also denotes the offspring of unintentional misalliances between dogs of different but known breeds. Although they will not be pure-bred they will have some predictable traits of behaviour and appearance. They may also benefit from what is known as hybrid vigour, where their mixed inheritance reduces the chance of recessive faults arising and often produces a more robust, long-lived dog.

Mongrel

This term refers to those dogs whose ancestry is unknown, and whose appearance defies any informed guess as to their origin. They can be any size, build or colour and have any coat type. The only thing mongrels have been selectively bred for is to roam, which is of course, the outstanding qualification of both parents. A lot of nonsense is talked about mongrels, that they are better, more intelligent, healthier etc. This is in the lap of the gods. Getting a mongrel is like buying a lottery ticket. Having said that, I once saw a 'Heinz 57' that was the original shapeless and shaggy dog. It actually pointed perfectly well.

Out-crossing, line-breeding and in-breeding

Any dog that is bred from will pass on exactly 50% of its genetic inheritance to its offspring. With each successive generation this influence or contribution is halved again, so that a grandparent can only have contributed 25% of a dog's make-up and a great-grandparent half of this amount, i.e. 12.5%. You can go down subsequent generations in a pedigree and, by halving yet again, may work out the genetic value of each ancestor for yourself. Breeding from entirely unrelated stock (where no individual dog appears more than once in the pedigree) is called **'out-crossing',** and helps obscure the presence of undesirable, recessive faults. Each mating will have been chosen to counterbalance perceived weaknesses in either or both parents, with the resulting progeny likely to tend towards the average.

However, breeders, in pursuit of excellence, have always tried to increase the influence of a particularly valued dog or bitch by mating its relations together. The degree of closeness in their respective pedigrees dictates whether we say the offspring are '**line bred**' or '**in-bred**'. There is not, to the best of my knowledge, any definition of this difference so I can only tell you of my interpretation.

If a specific ancestor appears in a pedigree as both a grandparent and a great-grandparent it will have contributed a total of 37.5% of the genetic inheritance. It will therefore influence the resulting offspring by more than any other grandparent (25%) or great-grandparent (12.5%). However, the contribution of that ancestor remains less than the degree inherited by a direct son or daughter (50%). Revisiting a desired animal in this way, is a means of slowing down the steady dilution of the original dog's qualities, and is called **'line breeding'**. The risk with this practice is that the original dog, to which you have line-bred, may have carried unseen and unsuspected faults in its make-up. As you increase the percentage of his or her influence, there is a danger that these may become more evident, just as the desirable qualities might. Supporters of the practice of 'line breeding' will claim that it is better to find out what faults exist within a bloodline than to hide them by continually breeding to unrelated individuals (out-crossing). The faults, in this latter practice, may remain a 'ticking time-bomb' within the breed.

Line breeding will highlight, or exaggerate, both good and bad points. Anyone embarking on this particular journey must be prepared for disasters as well as triumphs.

Sometimes a breeder will mate very closely related stock together. This might be father to daughter or son to mother, brother to sister, granddaughter to grandfather and so on. In this case the percentage of genetic influence (**consanguinity**) of one animal will be greater than 50%. When this occurs, the offspring will be **in-bred**. With so little genetic diversity the chance of any previously unsuspected, but undesirable, recessive weaknesses showing up is greatly increased. Despite this it is theoretically possible for in-bred animals to be more vigorous and healthy than average, but only if the basic stock does not contain inherent, recessive weaknesses. There is a famous family of laboratory rats that has been bred for many decades from a brother to sister mating. Their millions of descendants are used throughout the world for experiments as they are, to all intents and purposes, the same rat. These rats live longer, grow bigger, are more fertile and have fewer health problems than any other strain of rat.

Recessive and dominant traits

Before going any further, I should state that genetics are still a new and rapidly evolving science. I can only put down my own, extremely simplified understanding of the basics. This is done with a deliberately 'broad brush' in the knowledge that every day brings a fresh insight into a fiendishly complicated subject.

We know that each dog has a genetic DNA code that is made up of thirty-nine chromosomes. Each chromosome comprises many pairs of genes which, in turn, are composed of many different constituent parts, providing almost limitless permutations. For every characteristic, whether these are physical or temperamental, visible or unseen, these combinations of pairs of genes dictate how each individual dog is made. When the dog is bred from, it randomly passes one from each pair of its chromosomes on to its offspring, while its mate does the same. Each pup will therefore inherit a pair of genes for each characteristic, but one will have come from the sire and one from the mother. This means that both parents will have an equal influence on the genetic inheritance of the pup and its eventual make-up, as they can each only contribute to it to the same extent.

There is no guarantee that each puppy in a litter will receive exactly the same distribution and combination of genes, any more than human brothers or sisters are likely to be identical. The eventual appearance and temperamental characteristics of the individual puppies will vary within the litter. Experienced breeders welcome a 'level' litter, i.e. one in which there is no extreme diversity of type and size. From this they know that the parents are truly 'geno-typical' (refer to the second paragraph of this chapter), which is what they want. Conversely a nest of pups of disparate sizes, colours and types indicates that the genetic mix is still wildly variable.

What causes variation depends on the presence or absence of '**dominant**' genes. A very simple example of this was a black whippet dog called Wishy Was, that raced in the 1960s. He was used prolifically at stud, and was mated to bitches of every conceivable shade and hue, yet all his progeny were either black or blue (the dilute of black). So, it would be accurate to claim that he was dominant for this colour, and possessed no other genetic possibility for this characteristic within his own make-up. This is quite unusual in dogs, as black and red colouration are normally of fairly equal value, which can be seen in the mixed colours of brindle, black and tan, chocolate, liver, blue-fawn, tricolour and sable. Anyone at that time, who specifically wanted to breed black whippets, could confidently use Wishy Was and get what they sought.

However, many of his black or blue sons from bitches of different colours, would produce a mixed colour litter as they had inherited only half of their genetic make-up from the father. The unseen other half of their breeding was still present but invisible.

The opposite situation occurred some ten years earlier in Lakeland Terriers. Unexpectedly two liver and tan pups were born in an otherwise regulation litter. Although old versions of the breed standard quoted this as an acceptable colour within the breed, nobody had ever seen one. As it transpired, the two pups, one of each sex, were good specimens and won well at shows. In an attempt to capitalise on this good fortune the breeder embarked on some very close in-breeding indeed. The parents were mated back together, father was mated to daughter and mother to son, and eventually the two siblings were mated to each other. Despite this the colour never reappeared. However, it must still be there, lying dormant yet recessive in the breed, ready to spring up again at some future date.

Of course, most inheritance does not depend on just one gene, or even a single chromosome, but is often multi-factorial, the product of a combination of a variety of genes. Very quickly the permutations become endless. However, where a trait is commonly seen, it is probably dominant. Only in the absence of such a dominant gene will you become aware of any hidden recessives, and they can only appear when both parents pass on that recessive gene.

Fortunately, the qualities of health and vigour that we want to instil into our bloodlines are normally those which are genetically dominant. Those which are recessive are always more elusive, and can easily disappear with the introduction of dominant genes from a new sire or dam. They may still be there, although not evident, and can resurface in future generations. Where a desired virtue, whether of appearance or behaviour, depends on a mix of different genetic factors, then the task of 'fixing' them becomes frustratingly complex. Equally it can be as difficult to eradicate unwanted genetic factors.

Anyone contemplating breeding dogs must understand that any individual parent can only pass on those visible or hidden qualities it has itself inherited. When mated there is no guarantee that the actual combination of genes they pass on for each quality is the one which you sought. When unexpected traits appear, this is only possible because that combination of genes already existed, unsuspected, in the genetic make-up of its parents. Only where one parent has itself inherited two dominant genes for a single trait, can you be sure that it can only pass on that dominance to its offspring. The mating may have been selected to rectify a perceived weakness in the bitch. Quite probably the resulting pups, while not exhibiting the undesired trait themselves, will still have inherited the recessive, unwanted gene from the mother. This means that their own offspring can as easily pass on the original weakness, unless paired to a mate with the desired dominant qualities.

A lot of emphasis is placed on seeking a '**prepotent**' stud dog. This term describes those male dogs that are known to throw particular characteristics. To do this consistently implies that it has a combination of the desirable dominant genes in its make-up, and fewer unwanted recessives. However, no matter how consistently it sires dogs of a type that seems ideal, there is nothing to stop its resulting offspring passing on hidden, undesirable qualities inherited from their mother. Sadly, over-use of outstanding dogs, in numerically small breeds, has sometimes resulted in the introduction of severe health problems, which it will take many years to eradicate. Nowadays our modern knowledge of DNA testing enables us to identify 'carriers' much more readily than before.

While the genetic inheritance of a puppy is received in equal proportions from each of its parents, it is important to realise that the bitch has the greater influence as, to some extent, she determines the health and vigour of her pups. A brood bitch that whelps easily, and produces strong offspring that thrive, is of inestimable value. Her temperament will affect her behaviour and this, in turn, influences that of her litter. Selection of brood bitches relies on more than just prize cards and rosettes!

The genetic pool and resulting vigour

The total genetic pool of dogs of all breeds is vast, and contains the building blocks for every extreme of shape, colour, coat texture and behaviour that the domestic dog can exhibit. Each specialist breed has been selected to display only a narrow range of this total, exaggerating some qualities while excluding those inherited traits that are deemed undesirable. Geneticists stress that this narrowing of the gene pool may adversely affect the vigour and health of the breed. They can point to reductions in basic indicators of robustness, such as longevity, litter size, fertility, natural whelping and freedom from inherited health conditions.

In some breeds the defining traits may themselves be recessive ones, originally stumbled on by accident and retained by a systemic elimination of less desirable dominant ones. To fix such features requires a conscious reduction of the gene pool. It can also be that dominant genes, that are sought after, may carry unwanted side effects.

An example of this is the taillessness that can be found in Brittanys. This gene is dominant but also semi-lethal. An embryo receiving a double helping of this gene, i.e. from both parents, will not develop and will be re-absorbed. The only surviving embryos will either not have the dominant gene at all (and will have a full tail), or will have one dominant half for taillessness and one recessive half for being tailed (and will have no tail at all or a tail of reduced length). Mating two visibly tailless individuals will reward you with a reduction in litter size of approximately 25%. Statistically the surviving 75% of the litter will consist of 25% with tails (where both parents contributed a recessive 'tailed' gene) and 50% will have no tail or a shortened tail. These latter will carry both the recessive for being tailed (and so survived the womb), but also the dominant (visible) gene.

In the diagram below the dominant, tailless gene is expressed as 'T', while the recessive gene for a normal tail is indicated by 't'. Both parents will carry two genes but each parent will only pass on one of these each to their offspring. This allocation of a single gene from each parent is random, so will rarely result in the statistical distribution of the progeny as depicted. You could, theoretically, have all pups born tailless, all pups born tailed, or even no pups at all from a successful and fertile mating, because *all* the embryos inherited a double helping of the lethal dominant gene for taillessness.

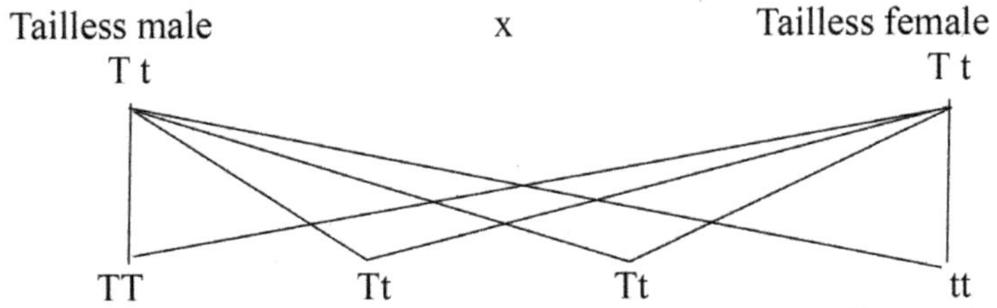

25% Doubly tailless absorbed in uterus

50% Tailless and able to produce tailless and tailed progeny

25% Tailed and can only produce tailless progeny if bred to tailless mate

Where one parent does *not* carry the gene for taillessness the ratio of tailed and tailless pups will be as depicted below. Because it is impossible for any embryo to receive a double dose of the dominant tailless gene (TT) the average litter size will be one third greater, as no embryos will be lost through re-absorption.

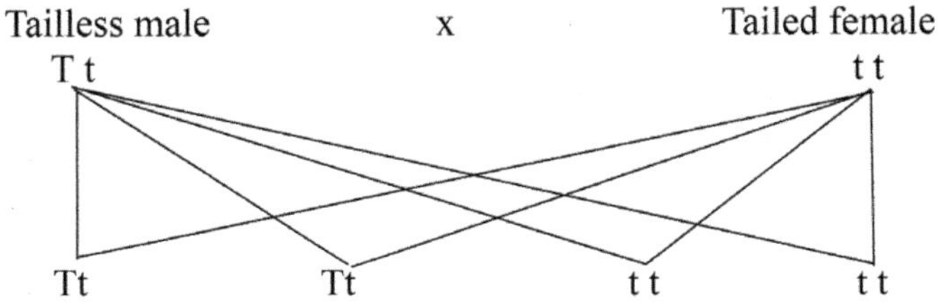

50% tailless and able to produce tailless progeny to tailed or tailless stock

50% tailed and can only produce tailless progeny if bred to tailless animal

Following the introduction of the Animal Welfare Act of 2007, when legal docking became increasingly scarce, many Brittany breeders actively sought tailless stock. This concentration on a single quality may have narrowed the breed's gene pool significantly, at the expense of other characteristics. Show enthusiasts only exhibit tailless specimens, and their breeding stock is mostly tailless. This has resulted in a reduction in litter size, and a drop in the numbers actually being produced throughout the breed.

Breeders are sometimes attracted to individual outstanding sires, and there is a chance that the breed's already narrow gene pool will be further diminished, as fewer genetically diverse dogs will be contributing to it. In a numerically small breed, and in the UK this applies to most HPRs, this is a very real danger. Nowadays the Kennel Club has created an accessible online breeder service which calculates the degree of consanguinity (in-breeding) of any projected mating. Any breeder should make use of this at the planning stage, before actually using their selected stud dog.

We have a responsibility to keep the genetic pool as wide as possible, by sourcing unfashionably bred breeding stock, or importing stud dogs from their country of origin, where a wider range of bloodlines will exist. This last course is a 'quick fix' that is fraught with hazard. The importer will probably have no knowledge of the wider family of the intended purchase, and without this there is a danger that he or she could be buying as many problems as benefits. It is also a fact of human nature that no breeder will want to sell their best animals.

The alternative is to study the wider families that exist within the breed in your own country. You are more likely to know of 'problem' relatives, and be able to better assess the *average* quality of the family you are buying into. Initially look for those bloodlines that consistently produce large litters, are breed-typical and have sound temperaments. Select from these the lines that are seen to have a high percentage of the qualities that you value. Be wary of (but do not necessarily exclude) top show lines – as success in the show ring often comes from exaggerating specific qualities. This can only be achieved by a narrowing of the genetic pool!

Quality control

Hindsight is a wonderful thing, and wisdom only comes with experience (and sometimes not even then). A good breeder will seek feedback from his puppy purchasers about health and behavioural problems they may have had with their pet. This is just as important as looking for show or field trial results. The majority of pups do not get to compete, so judging by these criteria is not necessarily representative of your stock. Some breeders can keep producing winners by sheer weight of numbers, while closer scrutiny may show that they equally produce some really poor physical specimens, or ones with no working ability and suspect temperaments. Every breeder will do this to some extent, so do not be too harsh with your judgement. Sometimes an experimental introduction of fresh 'blood' does not work, and the resulting litter does not move the breeder's long-term plans forward at all. But they will have learned something from the experience.

Objective measurement can be done by finding out what proportion of your stock changes hands (even ending up in rescue). This may reflect on the experience of the breeder in monitoring the prospective buyers thoroughly, and in advertising sufficiently so that he or she can select from a choice of suitable homes. But a dog that is a pleasure to have in your house will not be sold. Starting as a small breeder you may not have enough stock out there to give a fair comparison, but create a starting point and keep monitoring it.

Another way to measure is to check on what percentage of your stock ever gets bred from. Only about 3% of any breed contributes to the next generation (and of these only about one in five, or .06%, of the overall total is male!). The fact that owners choose to breed from their pets suggests that they believe that they are worth it. On checking the 269 direct descendants from my original foundation bitch I found nine bitches and, surprisingly, nine dogs had been bred from. This compares very favourably, with 6.7% contributing to the breed.

Chapter 11
The Practice of Dog Breeding

My parents were 'backyard' dog breeders, having the occasional litter from what was, essentially, a succession of family pets. This rather demeaning description of them is perhaps an unkind one as, like many other enthusiasts, they produced some excellent dogs in just this way. Done properly, the backyard breeder knows his or her stock thoroughly and can give very individual attention to them. Certainly, I never questioned if this was acceptable, until the nasty gremlin of ambition crept into my soul.

The breeder's ambition

Anyone thinking of breeding a litter of pups should have a clear idea of why they are embarking on this journey. The first requirement of any dog is that it should be pleasant and healthy. If your bitch does not fulfil these two basic criteria you should stop at this point. Experience tells us that dogs that are not pleasant will have to be exceptional in other ways if they are to be kept. On the other hand, an individual that is only moderately successful, but gives its owner affection, loyalty and trouble-free companionship, is more likely to be cherished. Equally a dog must be physically sound and robust enough to do its job. One that continually racks up vet's bills, or cannot do a proper day's work, will always be a burden.

Having convinced yourself that you have a good enough '**foundation bitch**', which you hope will become the matriarch of your family of dogs, you need to be certain of where your aspirations lie. In many breeds there is a huge gulf between those strains which work, and those which are winning in the show ring. With HPRs this difference is much less evident, and it is still possible to aspire to produce genuinely dual-purpose individuals that can triumph in both spheres. However, precious few dogs actually manage to do this. You may, therefore, be faced with a choice of directions. Provided you have an interest in one or the other, it makes sense to concentrate on the area you are familiar with, as you will be clearer about what qualities you are looking for. Be guided by the best qualities of your bitch, but choose a sire that complements these without detracting from them. In other words, he has to be at least her equal where her strong points are concerned, and her superior in those areas where she is weak.

I referred to 'your bitch' in the above paragraph, as it would be foolhardy to try to found a breeding dynasty on a dog. The bitch owner can, if he or she has deep-enough pockets and is prepared to travel anywhere, use the best dog in the world to sire the litter. The quality of your stock can therefore be improved rapidly. Conversely nobody will bring the best bitch in the world to an average stud dog, so any matings he has will be with lesser bitches, and their progeny will tend to be ordinary. It is a truism of breeding that stock will always tend towards the average. However, do not completely discount the value of breeding from lesser stock which is healthy and temperamentally sound. These average, unfashionably bred progeny may still be the salvation of a breed that has suffered from excessive use of a limited number of stud dogs, and has acquired inherited problems as a result.

Before setting sail on the stormy waters of dog breeding you must ask yourself the following questions, and be prepared to be honest in your answers.

1. **Do I want to be bothered with the extra work, worry and expense of breeding a litter?**
2. **Do I have the time and facilities to do this properly?**
3. **Is my bitch of a quality that will give me a reasonable expectation of selling the pups?**
4. **Have I analysed her qualities and shortcomings objectively, or asked a more experienced breeder to do this for me?**

If you have answered either of the first two questions with a negative, then stop right there. A 'No' to the third should see you abandoning any prospect of using your current bitch. Instead, you should consider if it might not be wiser to buy in a better foundation bitch. Persisting with a poor-quality bitch will cost you a couple of generations, and a lot of time and effort, before you reach what should have been your starting point.

The fourth question dictates if you will ever become a successful breeder in terms of contributing towards your breed's future health and happiness. There is nothing worse than 'kennel blindness', a malady that can afflict the obsessive breeder. Those who suffer from it are unable to perceive the faults in their stock and therefore make no effort to improve it. My own mother suffered from this (hopefully this too is not inherited!) and over a lifetime of breeding her own 'line' of dogs, never managed to breed a champion. It did not help that she was quite the worst judge of a puppy as well. Nevertheless, her many generations of dogs were robust and good-natured, so she did her chosen breed no harm.

A dog-breeding licence

In recent years (from 2018) the Government have been under pressure to control the activities of puppy farmers. These are breeders who basically mass-produce puppies in crowded, unclean conditions from poorly bred bitches. These bitches will be bred off on every season, until they cease to become regularly pregnant, when they are sold off as pets. To regulate this trade, and compete with illegally smuggled pups brought in from Eastern Europe, a licensing system has been introduced for dog breeders. This can be either a licence issued by your local authority, following inspection of your premises, or through membership of the Kennel Club's Assured Breeder Scheme. The former charge whatever amount the local authority dictates, while the costs of the latter are standard throughout the UK, but still quite expensive.

The intention is to regulate the standard of care that goes into the breeding of healthy dogs, and is to be applauded. Owners who want to breed off their pet may be exempt, if they do not breed more than a litter every two years or so. It is the regular breeders, perhaps breeding largely as a profitable venture, that are the object of this control. Check with your local council for costs and procedures, and compare with the Kennel Club scheme. It may often come down to a 'postcode lottery' as to which is the most economic route to take.

When to breed off a bitch

The Kennel Club has a code of practice for dog breeders that it enforces through its registration system. It will not register puppies from a bitch that has been mated before she is twelve months or after she is eight years old, nor will it register more than four litters from any one bitch. Some breed clubs have their own codes of conduct that are stricter than this, restricting further the age for breeding and the number of litters that is acceptable. This is not only done to prevent any exploitation of the bitch, but also to stop any bitch having too great an influence, for better or worse, on the breed. I have never understood this last point, as an outstanding bitch can contribute as much to the breed's gene pool as a dog, and nobody suggests we should restrict the number of litters he can be allowed to sire.

Until you know your bitch thoroughly you cannot accurately judge her good points and her weaknesses. Lacking this you cannot sensibly select a suitable mate for her. Time will help you be better informed and make wiser choices. For this reason, if for no other, it makes sense not to breed at the first available opportunity. Your breed will also recommend various health screening programmes, to identify carriers of hereditary diseases, or to measure the degree of hip dysplasia. Your bitch must be old enough to have had these done before being used for breeding. Check with your breed club's health representative (every club is required by the Kennel Club to appoint someone for this responsibility) and find out what tests are available or required.

Breeding on successive seasons is definitely not a good idea. The bitch is able to conceive, but should be in the best possible condition to feed pups in the womb, and after they have arrived. The recommended minimum gap of twelve months ensures she has time to rebuild her energy levels, and replace the minerals and hormones that her first litter depleted. Again, the Kennel Club will not accept registrations for puppies bred from a bitch within twelve months of a previous litter. This enforced delay gives you the opportunity to learn something of the qualities and failings of the first litter, and to take this into account before choosing a sire for the second. Breeding does not stop when you sell your puppies. Following up to get consumer feedback is a very necessary activity.

Choosing a sire

In accurately assessing your bitch's weaknesses and strengths you will have arrived at a 'job description' for potential mates. I find it quite useful to list the qualities I see as essential, and to fill in these boxes with my bitch in mind. Then I can use the same pro forma for any stud dog I am considering. It is surprising how this helps objective decision-making! Create your own list based on those qualities you think are the most important.

Selecting the sire should be done several weeks before your bitch is due in season. The next step is to contact the owner to check the dog's availability, and to ensure that he or she is happy to use the dog on your bitch. (Some stud-dog owners are unwilling to breed their dog with what they may view as sub-standard bitches, as he will probably be blamed exclusively for the subsequent substandard offspring.) At this time, you should establish the terms of the service. This will include not just the stud fee (which is usually a bit less or similar to the price of a puppy), but also if you can have a second mating to make sure. This is a wise precaution as any eggs produced by the bitch have a comparatively short lifespan, and a second mating doubles the chance of conception. You also need to clarify the conditions in the event of the service being unsuccessful. Will you get the choice of a full refund, or a return service when next your bitch is season?

Quality	Bitch	Stud 1	Stud 2	Stud 3
Health – hereditary defects	Good			
Conformation	Lacks substance			
Movement	Excellent			
Responsiveness	Average			
Obedience	Variable			
Initiative	Good			
Concentration	Excellent			
Hunting ability and work ethic	Excellent			
Scenting ability	Good			
Pointing	Excellent			
Retrieving	Indifferent			
Reaction to sound	Average			
Temperament	Excellent			

Sometimes a stud-dog owner will request a puppy in lieu of the stud fee. This is a great compliment as he or she is confirming that your bitch is of a quality that they appreciate, and would like to introduce into their own bloodlines.

Having taken all this advice on board you are now ready to join the strange band of fanatics that are dog breeders.

The logistics of breeding

When you decide to breed from your bitch you commit yourself to a chain of events that is unstoppable. Within sixty-three days she will produce a litter of puppies. These will require care and management. They will also need to be advertised, and you have to be available to answer enquiries and to welcome prospective buyers to your home. It will be another seven or eight weeks before you can start selling them, and they may take quite a lot longer to sell than you expected. Before mating your bitch, have a good look at your diary to ensure you will be on hand to perform your share of these responsibilities. Will a litter clash with holidays or Christmas? Does the shooting season coincide so that you will be without your companion and chief ally? Is there a better time to breed than others?

This last point is very relevant for breeders of working dogs, as few people want to buy a puppy during the shooting season. They are too busy with their sport, and anyway the short, dark days of winter are not conducive to toilet training etc. Most people like the long, warm, summer days to concentrate on a new pup. With luck it may be ready for a limited introduction to game by the autumn and, with another summer's training behind it, will give a good account of itself in its second winter. For this reason, most gun dogs are bred in the spring.

When to mate

Bitches normally come into **'season'** once every six months. Some small breeds have slightly more frequent seasons, while working dogs, because of their strenuous workload, may experience longer intervals. Cold weather or health problems may also delay the onset of a normal season. Successful matings can only take place successfully when the bitch is producing eggs or **'ovulating'**. This occurs over a four- or five-day window, approximately two thirds through the duration of the three-week season.

When your bitch comes into season or 'on heat' (**oestrus** is the correct term) you will see that her vulva swells and she will show a bloody discharge. The amount of blood will vary from bitch to bitch, and some fastidious bitches are so zealous at cleaning themselves you may actually not notice for a day or two. This is awkward as you really need to know how many days she has been in season, so that she can be mated at the optimum time. Therefore, keep a close eye on her and inspect her bedding daily for drops of blood when she gets up in the morning.

After about ten days this discharge will cease to be as bloody, and will become straw-coloured or clear. This change marks the start of the four or five days when you should expect your bitch to be most fertile. Nowadays your vet can take a smear from the vulva, and tell you how close to her peak fertility your bitch is. This is well worth doing, as travelling to a distant stud dog (and they are rarely close to your home) can be a time-consuming and costly business. An effective and practical test can be done if you have access to an experienced local stud dog of any breed. He will give clear indication of her readiness, as will her response to him, but you will not let a mating take place.

Advice on which day is ideal for mating is often misleading, as you may be unsure of when the season actually started. Generally, somewhere between the eleventh and fourteenth day is recommended. I have a personal preference for mating bitches later rather than sooner. My reason is that I suspect that this leads to a greater percentage of bitch puppies in the litter – which increases my selection if I am keeping one myself, but also because bitches usually command a higher price. My preferred day for mating, with these factors in mind, is the fifteenth or even sixteenth day! From fifteen pups, in my last two litters, only three dogs were born.

However I have known of a bitch that conceived on the first day of her season, and have owned a bitch that produced seven healthy puppies (five dogs and two bitches, in contradiction of my theory) after having been accidentally, and successfully, mated on the twenty-eighth day (although she also bred on the thirteenth day on another occasion). On another occasion, at a show, my experienced stud dog was exceedingly enamoured of a very successful show winner. The owner insisted her bitch was a couple of months out of season, but also told me that she refused to be mated. I felt that since both dog and bitch seemed very keen on the idea, we could have tried a mating and that it would have been successful. It was not put to the test, but I remain convinced that it would have worked.

These are exceptions and, for the novice breeder, the general recommendation remains valid. The bitch will help you by exhibiting obvious sexual behaviour. If you rub her on top of her pelvis, above the root of the tail, she should arch her back against your hand and, at the height of her season, turn her tail completely to one side. Your bitch is telling you she is ready.

Introducing the bitch to the stud dog

When you take her to the stud dog, which is the usual practice, do not expect her to be quite as forthcoming as she was in the safety of her own home. Faced with a strange environment and a rampant, rapacious dog, she may well clamp her tail firmly between her legs and be prepared to defend her honour with snapping teeth.

This might, in turn, put a sensitive dog off. For this reason, I would strongly advise you to use an experienced dog on a first-time **'maiden'** bitch. He will be less likely to take offence, and more prepared to persevere. Moreover, his owner will have more experience too, and that will help ensure things go well.

Allow her half an hour to get used to the surroundings – and the dog if he is steady enough to meet her, with both being kept on the lead

Handling your bitch

No matter how enthusiastic (or not) your bitch is she will be more relaxed if you are handling her. Your job is to make her stand while the dog does his job. The first time you take her to a dog she will probably spin round, and may even try to bite the dog from fear, and you may get a nip too if she panics, so do not let go.

You will need to keep a very firm grip on her head, by winding the lead round her jaws a couple of times and grasping her collar. It seems excessive with your friendly girl – but be prepared.

Maiden bitches also have an annoying habit of trying to sit down at the vital moment. Do not worry about this, as the back end is the stud-dog owner's responsibility. He or she will support her so that she cannot collapse, and may have to help steer the dog in if she is being unhelpful.

In this photo the bitch has previously been bred from, but the dog is a novice. Anyone hoping to introduce their intended stud dog to his duties is advised to seek an experienced brood bitch for his first time.

Given the chance to proceed at his own pace – without being snarled at or bitten – will ensure he is more confident next time.

Never leave them to sort it out for themselves! As soon as he shows genuine interest, keep a good hold of her.

Once the dog has mounted and penetrated the bitch they will become '**tied**' together. Eventually the dog will try to turn and face the other way. Firstly, he will dismount and just stand beside the bitch. Then he will lift one hind leg over her back and will eventually, in most matings, be standing looking in completely the other direction. This is perfectly normal and does not hurt either of them. The bitch has a sphincter inside her uterus that contracts and locks the dog in position. You must keep hold of her until she releases this. It can take quite a long time, but is usually between ten and twenty minutes. Experienced breeders always like to see dogs tie like this, but bitches can sometimes conceive without this happening, when it is said to be a **'slip mating'**. The black dog in these photos was from such a mating.

The mating has been accomplished and the dog and bitch will remain 'tied' together for some time. They must be held to prevent either from moving away too soon and injuring the other.

Having taken all this advice on board you are now ready to join the strange band of fanatics that are dog breeders.

Repeat matings

It is always worth arranging for a second mating, a day or two after the first. Each egg the bitch produces will only survive for about twenty-four hours. The dog's semen, however, will live for up to three days. This spreads the possibility of insemination over a longer period, and is therefore more likely to result in success. Do check that the stud-dog owner that is happy with this, and that the dog is available. There should be no extra charge for this. Very often the dog and bitch will both show a great deal more enthusiasm for the job on a return visit. In part this is because they already know each other, but often it is because you first brought her to the dog at an early stage of her season – she was not quite right.

Misalliances

Misalliances can happen (although they shouldn't), and bitches do sometimes escape and get mated by an unintended suitor. Equally the sire might be quite desirable but the timing of a litter may not fit into your own plans. In both cases your vet can give an injection, which will prevent the bitch remaining pregnant. It is the canine equivalent of the 'morning-after' pill. These drugs are very powerful and can affect the bitch's reproductive system for a very long time. You should certainly think twice before thinking of breeding from her at her next season. Of course, science is always making progress, but this is an area where you tamper at your peril.

If your bitch has already been mated to a dog of your choice, and she subsequently gets caught by the farm collie down the road, you may opt to let her pregnancy run its normal course. If you are lucky you may have a perfectly good litter, that is obviously of the breeding you planned. Or you may be horrified to see that some of the whelps look like the intended, pure-bred Weimaraners, but half have traditional black and white collie markings. This is called a **split litter**.

The implications of a split litter affect your ability to register the pure-bred puppies. With DNA testing you may now be able to prove their ancestry, and the Kennel Club may accept evidence of this and register the pure-bred ones. Your bitch will not be devalued by this experience, and any subsequent litters will not be tainted by her previous scandalous affair.

Paperwork

Once your bitch has been successfully mated, the stud-dog owner will require payment, unless he or she has agreed to wait until the puppies arrive. Personally, I like the fee at the time, as people can become remarkably forgetful. On paying this you should be given a receipt that states the terms, i.e. free return service or a full refund if the mating is unsuccessful. You should also receive a copy of the dog's pedigree, which you will need if you are to fill in your puppies' pedigrees accurately. The third document is a copy of the Kennel Club's 'Application to Register a Litter'. It can be downloaded from the Kennel Club website if the stud-dog owner does not have a copy. This form, which *must* be signed by the stud-dog owner, includes the sire's registered name and registration number, and confirms the date of the mating. Without it you will not be able to register your own pups.

Where your breed specifies health checks, you should be given a copy of the stud dog's certificates relating to these. This information, when added to the bitch's results, form a valuable source of knowledge for future purchasers.

Is she in whelp?

No matter how experienced you may be as a breeder you will always be eager to assure yourself that your bitch is in whelp. Some help you by going 'dreamy' from the outset. This change of temperament is quite noticeable. Others are more modern mothers, and just carry on as though nothing had happened.

Eight weeks in whelp – this bitch was carrying nine pups but I had known she was in whelp from the first day – as her behaviour softened instantly.

Despite your intention not to feed her abnormally, you will almost certainly find little opportunities to give your 'girl' extra treats and to spoil her a bit. This, and the fact that most bitches slow down after a season and retain more fluids, means that she will put on a bit of weight. You will be encouraged to believe that the mating was a success, but you cannot be sure until fairly late in the gestation period.

It is only in the last couple of weeks that the embryos will grow significantly in size. A healthy bitch does not need any major increase in diet, and should certainly not be allowed to become fat, which may affect her ability to whelp normally. She will have carried any puppies high up under her ribs at first, and it is this thickening and broadening of the rib cage that is often the first reliable sign. Eventually the litter moves back, through pressure on the available space, until her belly is taught and swollen. This will be most noticeable when she is lying down.

By the seventh or eighth week you will be able to feel small kicks in her side, and these can be clearly seen in smooth-coated breeds. Even then a bitch with only one or two whelps inside her may be little changed. I have worked in-whelp bitches to within two weeks of whelping. Predictably they did slow down and I reduced the workload accordingly. Of course, getting them over or under fences became a little more difficult!

Scanning

It is now quite usual to have a bitch scanned to see if she is in whelp. I have never understood the need for this. By the time the embryos can be spotted on a scan it is too late to do anything about it. You cannot send them back and neither can you make them appear if she is not pregnant. Scans are notoriously inaccurate and frequently miss pups or underestimate the number. (Recently a friend's bitch had nine pups despite a scan predicting six.) The vet, in this instance, was reluctant to dock the 'extra' pups, suspecting that these were unrelated stock being 'laundered' for legal docking. Only the threat of legal action persuaded a change of heart!

False or phantom pregnancies

Many bitches offer symptoms of being in whelp. This can happen even when they have not been mated. They will become dreamier and will retain fluids. This is quite normal and is often an indication that they are fairly maternal in nature. They may even, in extreme cases, produce some milk. After the time when they would normally have produced pups, they will return to their habitual shape and behaviour. A bitch that regularly has false pregnancies may well develop a septic womb (**pyometra**) in later life and need spaying.

Bitches that are unusually attractive to male dogs, even when they are not in season, may have a hormone imbalance due to a developing pyometra. If yours is one of these you should certainly have her examined prior to taking her to be mated.

Is she on time?

The gestation period is nine weeks (63 days), but it is not unusual for bitches to work to their own calendar. One litter, from an experienced brood bitch, arrived a full week early. They were all fully formed and a good weight, so I guess she knew when they were ready. Being this early is unusual but a **delay** of forty-eight hours beyond the due date is much more serious, and should see you calling the vet.

Health problems

Bitches in whelp are normally vigorous and healthy, but you can get problems. Very occasionally a successful mating may result in no puppies being born, because the bitch has contracted some minor infection that caused her to reabsorb the embryos at an early stage. If this happens later in the pregnancy the puppies may die inside her, but she will not be able to absorb them. Instead the puppies will begin to decompose and she will almost certainly suffer from a septic womb. The first indication that things are wrong, apart from your gut feeling that not everything is fine, may be a dark green or blackish discharge. Taking her temperature will probably confirm your worst fears. The best you can do is to get her to the vet's – fast. You will be faced with a caesarean operation to remove the dead pups, and may also have to have her spayed at the same time. If you are lucky this last operation may not be required, and she may still be able to breed again. NB I have heard of the loss of litters through decomposing dead embryos, but have never experienced this disaster myself, so do not worry unduly – it is a very rare exception.

Loss of a litter is distressing, but you must remember that not everything can live. Sometimes, despite the rude good health of both parents, there may be a genetic 'clash' that triggers re-absorption, or death, of the puppies. Some genetic combinations can be lethal, and this may be Nature's way of controlling the production of sub-standard individuals. I have only had two bitches visit my old stud dog that failed to produce live puppies to him. On inspecting their pedigrees, they were very closely related (and very in-bred, which is known to affect fertility). Was there something in the genetic mix that was never going to work? Certainly, if you have a bad experience with your bitch, you should consider a totally different stud dog for the next litter. This is a situation where you want the stud contract to allow for a refund of the fee rather than a return mating.

Worming the bitch

During the pregnancy you will need to begin a worming programme. Your bitch will almost certainly have worm eggs lying dormant in her system, which will become activated by the changes in her hormones. By controlling the worms she hosts, you will reduce the worm load that she can pass on to her pups. My bitches are wormed at three, five and seven weeks in whelp. This is important, not only for the sake of the bitch and the litter, but also because canine roundworms can invade humans. Children will be at risk of a variety of conditions if they contract roundworms from the family pet. Being totally selfish about it I also have a rooted objection to spending money feeding parasites.

Getting wormers from the vet is more expensive than buying proprietary medicines from the pet shop. Vets will tell you that their drugs are more effective and therefore represent better value. Last time I compared the ingredients and dose quantity of the two they were identical, so you will have to make your own mind up. Whichever you choose do read the instructions carefully and follow them religiously.

Latterly I have experimented with supplementing diatomaceous earth in my dogs' feed (1 teaspoonful per day daily for medium-sized dogs) which is claimed to effectively control worm infestation. In my last two litters neither the bitch nor the pups were wormy (in fact only a couple of puppy motions had any evidence of worms in them), while her previous litter was very heavily infested despite the same, pretty zealous, worming regime. It's not conclusive but these first results were encouraging.

A newly qualified vet told me the current thinking. To ensure complete freedom from worms, you would need to dose for worms every day for a month to be sure of breaking the cycle. This would be pretty expensive, and there is no guarantee that reinfection might not take place within days.

Anticipating the whelping

With nine weeks' notice you should have everything prepared, but sometimes your bitch will pre-empt your plans, and start matters off earlier than you expect. Therefore, have the 'maternity ward' ready some days in advance of the big event. You cannot be moving furniture about, and be assembling a whelping box, while she is dropping puppies under the stairs or out in the garden!

For purely selfish reasons I preferred to whelp my bitches in the house. It is warmer and more convenient than sitting up all night in an outdoor shed or kennel. Nowadays I have a better-organised whelping shed. This has heating and light, and enough room for a comfortable armchair for the 'midwife'! Whelpings usually happen at night and it can get quite cold – so I can doze, wrapped up in a warm blanket, until my attentions are needed.

You need to be on hand for any small crisis (such as a puppy getting stuck behind the bitch and at risk of being squashed), and can casually keep an eye on matters. It is also kinder to get the bitch used to her 'den' in advance. If she is settled and confident, and has her own smell on the surfaces, she will be much happier. She will like your company too.

There are robust cardboard whelping boxes of all sizes available on the internet. These are perfectly adequate and are cheap, so that you can just burn them after your litter has been sold. If you wish to have a permanent whelping box it should be manufactured from planed wood or smooth, exterior-grade plywood. For the sake of storage, it makes sense to build it in sections that can be bolted together. The size will be dictated by the breed you keep. The shortest dimension should allow the bitch to stretch out fully, which she will need to do to suckle her pups. The longer side should be about half as long again.

My version (see below) has a removable half-top that will help conserve the warmth for very young pups. It acts as a convenient shelf for storing newspapers (the dog breeder's disposable bed linen) but can be removed if you need good access – which you will during the actual whelping. Design the box so that the front is hinged halfway up, allowing you and the bitch easy access. By the time the litter is more mobile you may be grateful for being able to raise this to the full height, to contain the increasingly active litter. This whelping box will be the puppies' home for the first three weeks. After this they will become more energetic and will need extra space.

Being ready also implies that you have previously arranged for flexible time off, or have another 'midwife' on standby!

Have you:

1. **Notified your vet (and/or experienced dog-breeding friend) of the impending whelping?**
2. **Made sure that he or she is willing to dock the puppies within the first few days?**
3. **Put up your camp bed/sleeping bag?**
4. **Have you had plenty of rest and early nights – to give you an energy reserve?**
5. **Has the car been filled with fuel – if you need to get to the vet?**

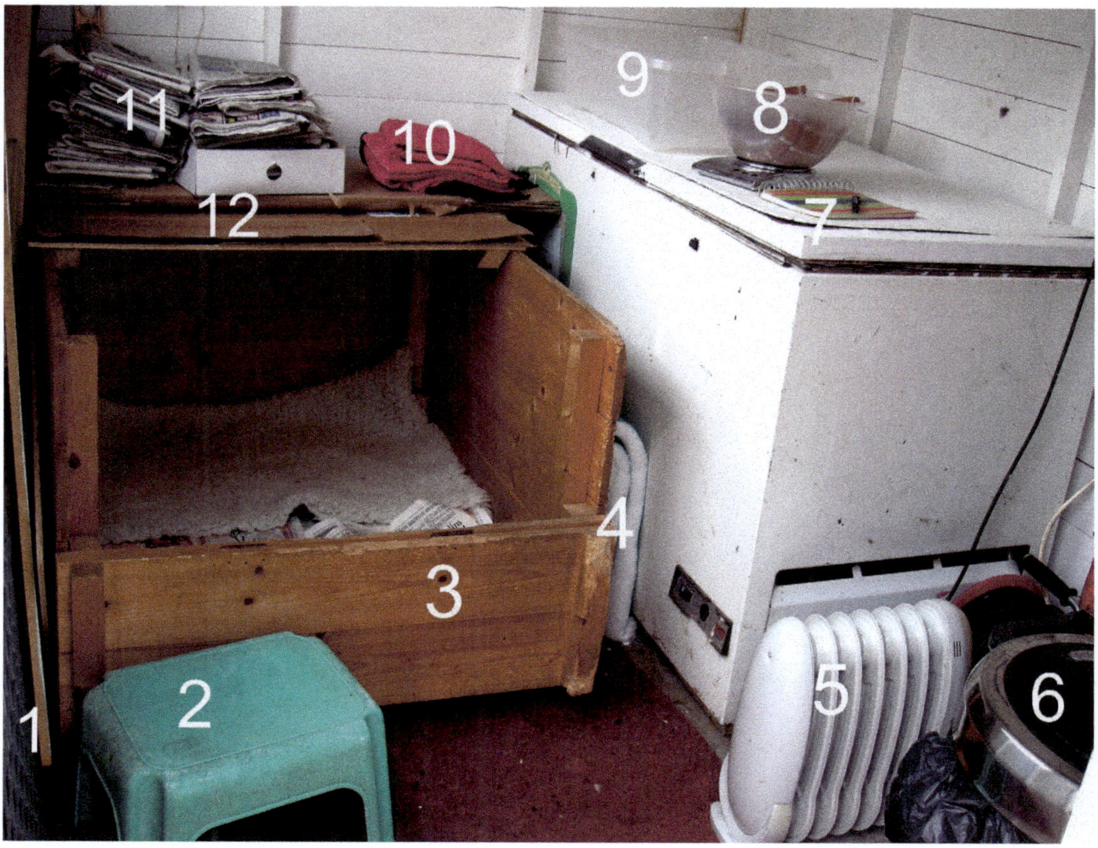

Make sure the following items are within easy reach during, and immediately after, the whelping:

1. A mesh inner door, so that the whelping shed is secure but the bitch can get plenty of fresh air
2. Something low and comfortable for you to sit on
3. A whelping box with a folding front will give you better access during whelping and can be raised as the pups get more mobile.
4. A supply of clean Vetbed for when the whelping is completed
5. Supplementary heating
6. A receptacle for soiled newspapers (whelping can be a soggy, messy business)
7. A notebook and pen to note times, sex, colours etc.
8. A stout box or basket to put pups in (the early born pups can get trampled on as she delivers their later brothers and sisters)
9. A supply of clean, old towels
10. A supply of old newspapers, giving extra insulation on top of the removable roof
11. A supply of flattened cardboard to re-line the base of the whelping box

Also: Very mild disinfectant and sharp scissors if you need to snip umbilical cords

The HPR breeds are, usually straightforward to whelp. They should be energetic animals with good stamina, and they do not have the exaggerated heads which cause difficulties in some breeds. Therefore, you can be fairly relaxed about things. As a precaution telephone your vet's, a day or two before she is due, to let them know– just in case.

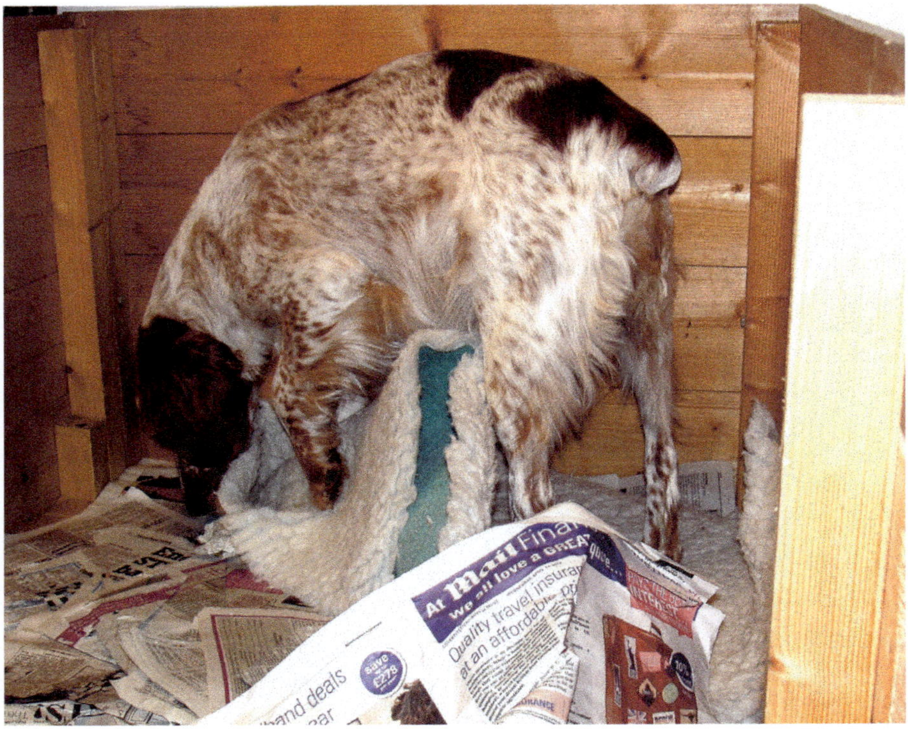

As matters progress, she will normally start scratching up the newspapers you have lined her box with. These should have been laid on top of some thick, flattened cardboard boxes. Scratching helps distend her pelvis and gets her stomach muscles warmed up.

The onset of labour

If you are observant you may spot that the bitches 'lump' has moved back a bit, and that her ribs look slightly narrower in consequence. She will be 'bagging up' underneath and her teats will become more enlarged, and she may show some loss of weight or muscle tone over her pelvic area. The first indications you will get, that 'things are on the move', will be an increased restlessness in the bitch. She will become fidgety and be unable to settle, and this stage may vary from as much as twenty-four hours to as little as thirty minutes (apologies for not being more precise). She may go off her food (although with greedy bitches this is not necessarily a reliable signal either) and will tend to want attention. It is advisable to encourage her to get into the whelping box, as she may otherwise select a completely unsuitable place for her labour.

Labour and the first puppy

Eventually she will strain properly and may show signs of fear or distress. **It is worth noting the time when this first contraction happens, so that you know how long she has been in labour.** If there are complications your vet will need this information. Some maiden bitches are traumatised by whelping, and you can only offer sympathy and comfort.

Labour is not protracted and, once the first puppy is on its way, you will not have long to wait. There will be a mighty heave and possibly a yelp from the mother. In the perverse way of dogs, she will almost certainly be turning round so that you cannot see what is happening. However, a dark or silvery blob will be seen emerging from the vulva. This is the fluid-filled envelope (**amniotic sac**) in which the pup has been living. With another couple of heaves the whole thing will slide out on to the papers.

As she may not know what to do, you must take over from the mother at this point – she will let you. Quickly prise the envelope away from round the puppy's mouth and muzzle so that it can breathe. The envelope is a very weak membrane and breaks open easily.

The pup will be damp and possibly may appear lifeless. Rub it vigorously with a clean old towel, from the pile you should have available. Almost immediately it will squeak quite powerfully, which will attract the mother. She should investigate, and then start licking it dry in a fairly robust way.

The umbilical cord

During these few minutes some pups may still be attached to the mother by an umbilical cord, with others this may break naturally in delivery. As nature and instinct takes over, she will chew this through. Some bitches are over-zealous at doing this and can pull the cord or chew it too close to the stomach wall.

This can cause an **'umbilical hernia',** so it is worth trying to pinch the base of the cord between thumb and finger while she is doing this. Such a hernia is quite common and you will notice, as the puppy gets a few weeks older, that it has a small soft lump protruding from its belly. What has happened is that the bitch, in chewing the umbilical cord too close to the puppy, has damaged the outer stomach wall, enabling the intestines to bulge into the gap. You may be able to push this back in with light pressure from your fingertip and, with luck, it may stay in – and the stomach wall will quickly grow over the opening. At other times the opening is too large and the puppy will eventually (at 12 months +) need minor surgery to sew the opening together. In this latter case you may need to offer the purchaser reimbursement, on production of the vet's bill.

Safely delivered – now she will lick the puppy dry

The placenta

A short while after the birth she will have another, slighter strain and will pass the placenta or afterbirth. This is fairly disgusting and looks not unlike a piece of rotten liver. Never mind, it contains vital minerals and vitamins for the bitch, and she will eat it voraciously. All this can happen so swiftly, while you are busy drying the new puppy, that you may even be unaware of the bitch devouring the placenta at all. So do not worry unduly if you lose count of the number of placentas delivered. Now is the time to change all the sodden, messy paper for fresh.

Further births

Mum will be fairly obsessed with her new arrival but pretty soon another one will be on its way. If she starts to strain again you will need to put the first born somewhere safe, as she may tread on it in her exertions. I like to have an old laundry basket beside the whelping box, lined with 'Vetbed', to pop the new arrival in when this happens. It will be perfectly safe under the towel which, until now, has been warming on the radiator.

The second pup will probably appear within fifteen minutes of the first. It is being pushed out by the pressure of the queue forming behind it, and the bitch is still strong and fresh. Subsequent deliveries will come at irregular intervals, averaging perhaps an hour or more apart.

With a large litter the bitch can become exhausted and may not get on with the job. If more than three hours have passed without a birth, yet the bitch is still restless and panting, you should call the vet. She may be suffering from **uterine inertia**, where the whole internal production line has just become tired and weak. If this is the case, your vet will administer an injection that stimulates the system and she should resume activities.

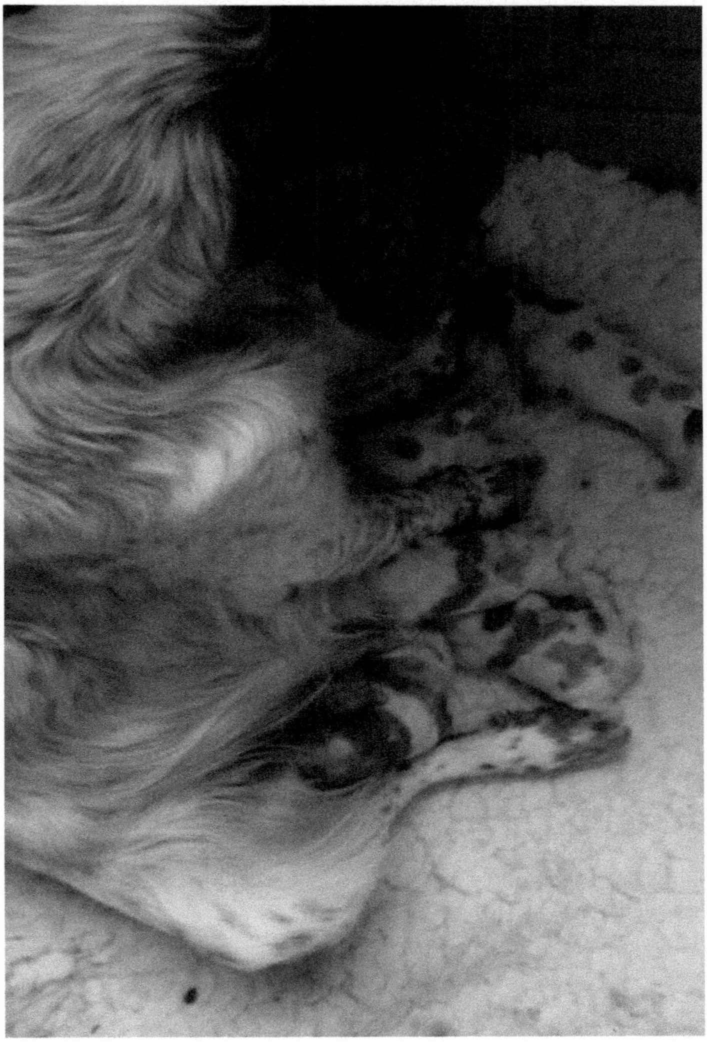

There are another seven pups to come in this litter, but the good mother will want the newborn pups with her between each birth. Note the pup at the top is already supporting its own weight, and can stand up, briefly, within hours of being born.

Between births put all the pups in with her, so that she can smell them and start suckling them. It is important that you make sure every pup does drink from her at this stage. Her milk gives them vital warmth but this first milk, or **colostrum,** passes on much of the mother's immunity to disease, and is important to the future health and vigour of the litter.

It is a good idea to weigh the pups at this stage and to record their weights, sexes and identifying features. This will subsequently help you to spot if one is not putting on weight, which may be a symptom of something being wrong.

When the last pup has been whelped the bitch will normally settle down in a blissful contentment, moving only to lick bottoms at regular intervals. Her breathing will become deep and peaceful. She may like a warm drink of milk. The soiled papers can be removed and clean Vetbed put down to keep her and the pups clean and warm. At this time she will be happy for the immediate family to come and look at the pups, but she will be stressed if all your friends and neighbours are invited round to see the new arrivals too.

Relaxed and happy to have a few visitors now the litter is complete.

Complications

The normal presentation of a puppy at birth is head first. It should have its head lying on its front feet and be facing downwards. You will not see this until it is being born, and the delivery is surprisingly swift. The vast majority of whelpings fall into this category.

If the bitch strains more than half a dozen times and produces nothing, then you may have to help.

Sometimes a pup will be facing the wrong way (it is called a breech birth), and it will be slow to appear. If you probe gently into the vulva, you will feel either the tip of the tail, or this and both back feet. In the latter case you should be able to grasp them gently, and ease the pup out with each strain from the bitch. Just let her push, and then hold the pup to stop it going back up again. A presentation like this is so common as to almost be considered normal, and does not take more than a few contractions to achieve success. However, if you cannot feel the feet because they are placed forward towards the head, the hips will be bunched up and will offer more resistance. In this instance you may need to explore inside the vulva with one finger, until you can hook each hind leg back in turn. Then you can hold the feet as described before.

Occasionally you will have a large pup that gets stuck in delivery. Because of its size its head can come out, but the front legs have been forced back, making the shoulders very bulky. Pulling on the head is dangerous unless done very gently, and it may not be possible to hook the front legs forward. Clearing the nose of the puppy from the membrane, so that it can get air, does not necessarily work as the ribcage may be compressing it so that it may be unable to breathe. Get swiftly on the phone to the vet in this instance.

A puppy that takes a long time to be born, and appears lifeless, can usually be resuscitated by rubbing it vigorously and keeping it warm. If it appears to have a lot of liquid in its lungs you can blow very gently into its mouth, and swing it up and down, from shoulder height to vertically down, several times. This should help drain any excess fluid. Administering a tiny drop of brandy (other spirits will do, of course) on a failing puppy's tongue sometimes gives them a bit of a 'kick start'. Be prepared for the fact that you cannot save every puppy.

Stages of development

Newborn pups are surprisingly mobile and will 'swim' around in small circles, whining plaintively if they become isolated from Mum and the comfort of their brothers and sisters. They quickly find her again and force their way in amongst the others. It becomes habit to keep counting pups, as one may be stuck behind the bitch and unable to get to food. Don't worry too much as they have lusty lungs and can let her know that they are complaining.

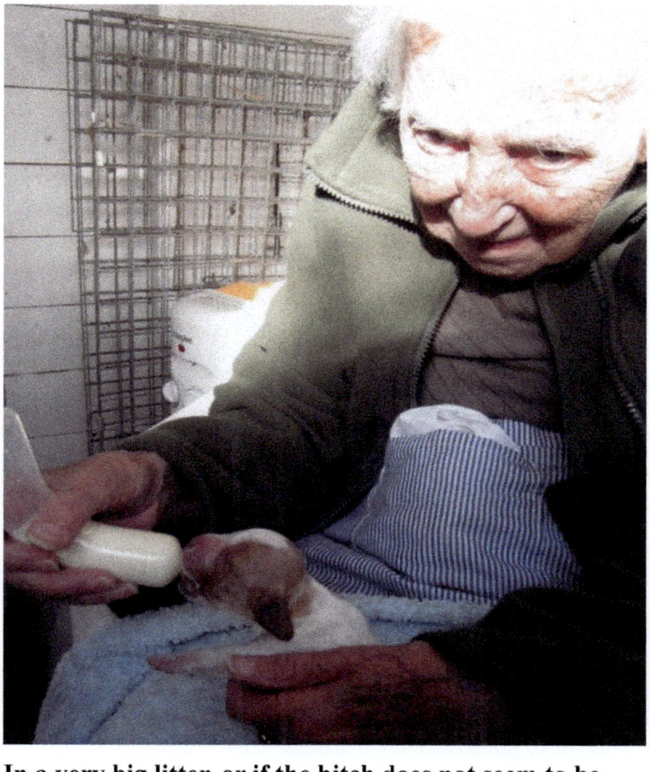

In a very big litter, or if the bitch does not seem to be producing much milk, you may be forced to supplement the puppies' diet with bottle-feeding.

Some new mothers suffer from eclampsia, and need an injection of calcium gluconate to help stimulate the production of milk. Pet stores do sell 'first milk', i.e. the one containing colostrum for newborn pups. After a couple of days, they can be moved on to 'first formula' baby milk.

Any small individual, that seems noisier than the others, may not be getting enough to eat. Make sure it gets regular time on the teat, and do not let the others push it off. This is particularly likely to happen in a large litter, where competition is greatest. In the wild the smallest would probably die for lack of opportunity but you can help it, and even supplement the mother's diet to make sure she has enough milk to go round. She should be fed ad lib with the emphasis on such foods as milk, eggs, porridge etc.

A litter of pups is one of the biggest time-wasters you will ever find. The many hours spent leaning over the side of the whelping box are the best paybacks for all the time and care you lavish on your dogs. From the first day you will see some whelps that are more active or determined than others. One, usually the biggest fattest male, is always on the back (largest) teat. Even if you swap him over to give a smaller sibling a better start in life, you can be sure he will back in the best place before very long.

Docking

If you are intending to dock your pups, this must be done in the first five days. Part of your preparation will have been to locate a vet who is prepared to dock, and then to inform him or her of the arrival of the litter and the number of pups. Probably you will have to travel to him for this minor operation, which will involve the bitch going too. Although she will be stressed by this, she will be much happier if she can accompany her pups. The vet should also want to check that she is OK as well. Refer to Chapter 1 on 'The Breeds' where this is covered.

Growing

For the first two weeks little will happen. The pups should grow steadily – which you should monitor through your routine recorded, weekly weighings. Regularly handling them gets them familiar with your smell and touch, and will accelerate their eventual socialisation. Your bitch will be perfectly tolerant of this.

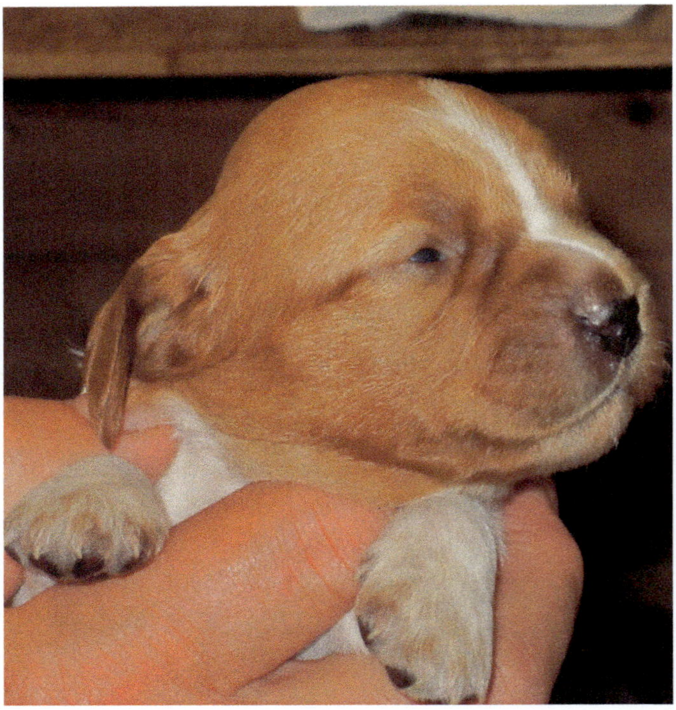

At about ten to fourteen days (individuals will differ) you may see tiny slits in the corner of the eyes.

Within another day or two both eyes will be fully open but the pups will be unable to see anything clearly. They will however become noticeably more active and may even begin to raise themselves up on to very wobbly legs. You have about seven days' grace before they start focusing properly and demanding more attention and space. It is about this time that pups start to be aware of sound, so speaking to them does have a value, as it accustoms them to your voice.

Weaning

Your pups will be dependent on their mother for the first three weeks of their life, by which age their eyes will be open and they can focus reasonably well. You will also be able to feel the first of their baby teeth coming through. Now is a good time to introduce extra food.

Start with offering a shallow dish or saucer of warm baby milk, with just one puppy sitting on your towelled lap. (At first most of it will go over you.) Try to get the pup to suck your fingers, where you have dipped them in the milk. Quite possibly it will show little interest in this substitute offering. Persevere over a few days, and the pups (which are growing and are therefore hungrier) will become more willing to try it. The addition of a soaked Farley's Rusk or a Weetabix may make the food much more palatable. Once this is being taken you can introduce some raw beef mince, mixed with warm, soaked puppy meal, as an alternative.

The whelping box outgrown – time to provide more space so they can develop motor skills.

Very quickly the pups will relish their new diet, and you will be able to feed them as a group. This can be bedlam, and it is quite a trick to manage to get the dishes safely on the ground before they are overturned by the surge of hungry pups.

Expect them to walk in their food, and fall head first into it as well. The size of dishes you use should enable about four pups at a time to eat. This may mean two or three separate dishes, but there will be less chance of the weaker or smaller individuals not getting their fair share. As the litter grows, the quantities of food will also have to be increased. In part this will be because the bitch will be producing less milk, and the fact that the pups are then taking less from her will help her 'dry up'.

By the time they are ready for sale they should be on four meals a day. These consist of:

1. a breakfast of warm baby milk mixed with soaked Weetabix (about one per puppy is reasonable).
2. lunch of raw beef mince, mixed with soaked Weetabix or puppy meal (about an ounce of meat and one Weetabix per puppy).
3. a tea-time repeat of the lunch.
4. a supper-time repeat of breakfast.

Bigger breeds will be taking larger quantities and the criteria is that, while they are not whining for food in advance, they should readily finish the complete ration at every meal. Once they are in their new homes they will probably want less, as they are no longer stimulated by competition with their siblings and are probably not burning off as much energy.

I usually recommend that new owners soon stop the late-night ‘supper’. This will speed toilet training. Each purchaser should be given a diet sheet, so that they are aware of how to increase quantities as the puppy grows, but decrease the frequency of meals.

A healthy litter needs space to run about and grow. Here Mum is avoiding having to suckle them, but hopes to get some leftover food – and half-brother takes advantage of this too. This interaction between the generations will give you a good idea of the family temperament.

Worming the litter

Once you have succeeded in getting the puppies to take food, you can implement a worming programme. Most prescribed or proprietary wormers are safe to use on pups from three weeks onwards. They will tell you the frequency and the dosage to use, and these instructions must be followed. I expect to worm a litter weekly from three weeks and, instead of weighing and dosing each pup individually, I simply total the weight of the entire litter, and add the correct dosage for this to their feed. The bigger pups take more and the smaller ones less – so it is quite accurate.

After worming you should keep a very close watch on the litter. One with a heavy worm burden may have quite a dramatic reaction, both passing and vomiting bundles of worms. Clean these up immediately and dispose of them by burning if you are able. Otherwise place the soiled paper, with the worm-riddled motions, into a sealed bin liner for the dustmen to collect. Subsequent regular worming will see a reduction in evidence, until the pups are as worm-free as you can get them.

We are very lucky to have good wormers available to us. It used to be quite normal to lose a good part of a litter through worm infestation. Breeders before the Second World War would administer tiny doses of arsenic to their pups, and felt the inevitable losses were worthwhile, as the survivors were healthier for being worm-free.

Advertising

Do not wait until your litter is ready to go before advertising them for sale. There is nothing worse than being stuck with rapidly growing puppies that are in need of socialising and training. Without this they will fail to develop properly into pleasant, people-orientated companions, and their future desirability is already being jeopardised. HPRs are mostly fairly unusual, and they are certainly not the kind of breeds that you can sell to any buyer who comes along. Because of their 'high-maintenance' nature, you must really make sure that their future owners fully understand the degree of commitment they will need to provide. Otherwise you will be creating problems for the owners and the pups, which may result in dogs changing hands to increasingly unsuitable homes, or even being abandoned or put down. Therefore, advertising them in the local paper or on internet sites is not the best route and should be avoided.

Begin your advertising campaign by alerting your breed club, and any fellow owners you know, to the intended mating. They may run a classified section on the club website that sources buyers who are already interested in your breed. If you intend to breed on more than one occasion, you should seriously consider creating your own website – and linking it to the club site. The owners of the sire you have used will also be good agents. As experienced breeders themselves they will receive all sorts of enquiries, and would rather promote pups by their own dog than the produce of other kennels.

Within a day or two of the litter arriving you should tell all of the above the numbers, colours and sexes that you have. If you are a member of a gun-dog training society you should ask them to publicise the puppies through their newsletter and website. Lastly the dog and sporting press are powerful outlets and are used by those potential owners who have a serious approach to dog ownership, and are seeking well-bred and reared youngsters.

Registration

When you register the litter, which can be done online to the Kennel Club's Registration Department, you will have the additional option of putting your pups on their Puppy Register. Again, it reaches serious buyers. This is a persuasive argument for registering the litter promptly, rather than leaving it until the last minute. All pups should be registered before they are sold.

When you receive the registration documents back from the Kennel Club, you will also be given free veterinary insurance to cover them for the first four weeks after their sale. Other items can include vouchers for free food samples.

The registration document must be handed to the new owners when you sell the puppy to them. You will need to fill in the relevant details and sign it, so that they are able to transfer its ownership to their own name. A lot of people never bother to do this and I find I am, theoretically, the owner of some fifty-odd dogs. Recently I was contacted by a veterinary practice that had had an injured, lost dog brought in. They had scanned it for a microchip and found it was registered in my name. Fortunately, I keep good records and was able to contact the owner, who was frantically looking for his dog, which had run off after having been attacked by a Rottweiler.

Kennel Affix

Most breeders will have a registered Kennel Affix. If you intend to breed dogs on more than one occasion you will find an affix is worth investing in (it is registered through the Kennel Club). Nobody else can use this, and it will become identified with your style of dog. Every puppy you produce will be registered with your own particular 'trade mark'. For example, if your affix is 'Windbreak' then all your own dogs will have names such as 'Windbreak Beauty' or 'Windbreak Freddie'.

If you buy in a dog from another kennel it will carry their affix, but you can add your own as a suffix (i.e. it goes on the end of the name). So, if you buy a dog registered as 'Jobsworth Handyman' you can then re-register it as 'Jobsworth Handyman at Windbreak'. Look back through your own dog's pedigree, and you will see many instances of both types of names. Adding 'Windbreak' also tells buyers that you have acknowledged that dog as meeting your own criteria.

Managing a stud dog

Half of every dog comes from the father, and there are some who would argue that his contribution is greater. There is no scientific justification for this, but it emphasises that breeders do expect the stud dog to remedy all the shortcomings of their bitch. In a way this is sometimes possible because it is relatively easy to judge the qualities of an established male. He will probably already have sired a lot of progeny from a variety of bitches. Should he be perceived to 'stamp' his type, then you can be more confident in choosing him than you might be with another, less proven dog. Bitches can also be just as dominant, but they will produce fewer offspring, and only the breeder is likely to know how consistent she is as a dam.

Should you have a stud dog

Owning a stud dog can be time-consuming. Your dog will probably not be used very much unless he is a prolific winner in his chosen field. In fact, only 3% of any breed will be bred from, and of this small number, probably only about one in five (0.6%) will be males. This means that dog breeders work with a very small gene pool. If you are willing to let your dog contribute to this, even if he only sires one or two litters, you will be adding to the mix. Sadly, the majority of male dogs are now castrated, almost as a matter of course, by the veterinary profession. These are the same professionals who banned tail docking on the grounds that it is needless mutilation. Given a choice you might have your own views on which bits you would prefer to have lopped off!

Owning a stud dog appears like easy money. Think again. Before a dog will be used, he will have had to gain a reputation in either the show ring, at field trials or both. It is not easy or cheap to campaign a dog, and only the best few will ever be used. Even if you are the proud possessor of such an animal, you may find he is reluctant to perform with the ladies. Quite possibly he will have been corrected for overt sexual behaviour, such as riding people's legs and incessantly bitching. Faced with an unco-operative first mate, and with memories of past scoldings in his mind, he may lack confidence. The more you encourage him the more anxious he may become, and there are many dogs which could have been used at stud but never plucked up the courage to do it.

Experienced breeders that keep male dogs (and having dogs and bitches together at the same time is *very* hard work) will virtually train him in his duties from day one. When sexual behaviour starts, and it can be seen in dog pups of just three or four months, they are quick to encourage it. The pup will not be told off or slapped down. Instead the owner will make a fuss of it. They will also, from time to time, make a practice of handling its genitals, so that it is not inhibited should some intervention become necessary when being used at stud. Very often an over-keen young dog will benefit from its handler being able to steer it in, particularly if the bitch is not being steady, or is of height that makes a totally natural mating difficult.

When should a dog start his stud career?

This will vary according to limited opportunities, from breed to breed and between individuals. Firstly, the young dog should have been cocking his leg consistently for some time. This advertises that he feels mature enough to mark his own territory and is a signal that, mentally, he is moving out of the junior society and is prepared to assert himself among the big boys. Some males do this very early while still physically in puppyhood.

Just as importantly you must consider the dog's strength and health. Coupling with a bitch does put a strain on the dog's back legs, and hip flexor muscles. If you feel he is strong enough to jump fences he will be strong enough to do stud work. By and large small breeds, and more lightly built individuals, will be ready sooner than heavier, more substantial dogs.

A friend of mine who bred toy dogs, once brought her in-season bitch and her stud dog to our house. He was not very assertive and, unfortunately, the bitch was refusing to have anything to do with him. All she wanted to do was to flirt with a very young dog pup she had. After failing to have any success with the older dog I asked if it would matter if the young dog was used. Her reply was that he was only five months old, but that she actually intended mating him to this bitch in future. We decided to try this very young baby, and got an easy mating which resulted in a litter of five. Of course, toy dogs mature much sooner than HPRs.

Once you are happy that the dog is mentally and physically robust, you should work hard to try and get him an experienced mate. With one successful mating under his belt he will be much more confident – but may become a real pain socially. Having had his first 'sniff of the barmaid's apron' he will have a fairly one-track mind to start with. Eventually he should settle down, and an experienced dog can become very sensible around bitches, and will not bother them until they are ready to be mated. As I write this, I have a ten-month-old bitch in season, and my ten-year-old stud dog is lying beside her. Each day he checks her out and, in a couple of days' time, I will have to separate them totally, as she is still too young to breed from. Meanwhile they are never left together without supervision.

Having used your dog at stud you will need to be extra careful about security. He will be likely to roam if any similar tantalising scent is borne to him on the breeze.

Advertising a stud dog

Use the club website or, if your breed is sufficiently numerous to justify this expense, the dog press and some of the canine stud websites. Best of all have your own website with links to the club one. This will enable you to put out much more information about your dog, and provide lots of photos. Make a practice of taking his 'business card' with you to doggy events. I just have copies of my dog's pedigree, with his Kennel Club Hip Dysplasia certificate printed on the reverse, as this is basically the information people want when choosing a stud.

First contact

This begins with the telephone call from the owner of the bitch. Sometimes they are talking about a date far in the future, but usually their bitch has just come into season, and they need to arrange a mating in the next ten or twelve days. Not infrequently they are phoning because they want a mating now – talk about forward planning! If your dog is advertised at stud you need to make him available, and it is surprising how often a bitch can only be mated at weekends! A lot of bitch owners do not know what day she actually commenced her season, so there is also an element of guesswork in it. Ask them to describe how the bitch behaves when her back is rubbed etc. This will give you some idea of whether or not she is ready.

There is nothing more annoying than a bitch travelling a long way to your boy, only to find out that she is either not ready or, worse, that she has 'gone over'. Your dog will try, but may be met by snapping teeth or great timidity.

Introducing the 'happy couple' and briefing the owner

It is always best to allow the bitch some time to get used to her surroundings before you bring the dog in to her. When you do introduce them, this should take place with both of them on the lead. Try to let them get to know each other, so that she is a little more relaxed. This gives you time to assess how useful, or otherwise, her owner is likely to be. Many do not realise that they may actually have to hold their darling, while she is being forcefully deflowered. You will need to describe what they have to do in advance, because you do not want the bitch whipping round just as the dog enters her. Either in case she bites him, or in case she causes damage to a sensitive and vital part of his apparatus.

This is also a good time to do any trimming of surplus hair from around the bitch's vulva. There are times when you need to be able to see what is going on – is he aiming too high or is she too tall? You should already have done a similar job on your dog's underside. This quick tidy-up may also reveal that the bitch has almost dried up. The wise stud dog owner will have a tin of Vaseline handy to lubricate this area and ease entry for the dog.

If you are sensible it is worth having a cloth, soaked in mild disinfectant, to wash the dog's vital parts down after the two have separated. Usually he will have retracted his penis on being released by her, but if it is still extended it is worth giving it a quick wash-down. There is always the risk of infection and anything you can do to minimise this should be carried out.

Where to mate

Dogs do get into routines, and your dog will get used to performing in a given place. Indoors or out is the first decision. In many ways doing it outside is simpler, but remember that bitches may come for mating in the depths of winter, when it will be both wet and cold. For this reason, I always allow matings to take place indoors. When the dog and bitch are tied you will still need to be in very close attendance, and may well find yourself supporting the bitch to prevent her collapsing and dragging the dog down. To do this you will need to be on your knees, or sitting on the floor. Carpet is much better than wet ground!

It is understandable that big stud dogs will attract small bitches, and vice versa, as breeders try to establish standard sizes. Do not assume that a natural mating will always be possible. Using the great outdoors has its advantages, if the bitch proves to be an awkward height for your dog. We had a rockery in one garden where we could always get a mating, because the uneven ground made it possible for the two dogs to work out a suitable place to compensate for their height differences.

With a tall bitch, which you are trying to get mated indoors, you may need to straddle her hind legs wide apart, so that she is brought down to the stud dog's level. Alternatively, you can try to provide him with a step of some sorts to raise him up. I find the former is much easier. Small bitches can just be lifted up to the appropriate height, by placing your arm under her tummy.

Second matings

Offering a second mating, either on the same day as the first mating or a day or two later, will increase the chances of the bitch being impregnated. A healthy stud dog can provide a second mating within an hour or two of the first. For bitch owners who have travelled a long way this is a reassuring practice.

Terms of stud service

As a stud dog owner, you should supply:

1. **an <u>accurate</u> copy of your dog's pedigree**
2. **details of any veterinary screening it has had, e.g. hip scoring or eye testing**
3. **a signed and completed Kennel Club 'Application to Register a Litter' form (which can be obtained online from www.thekennelclub.org.uk). This records all your dog's details, the date of mating and must have your signature on it, so that the puppies can be registered before sale.**
4. **a receipt for the stud fee, with the terms of the transaction included in it. These details must record the name of the bitch, and of the dog, the date of the mating, your own and the bitch owner's names and addresses, and the amount of the stud fee, or the sex and priority of any pup you choose instead of a cash fee (e.g. second choice of the bitch puppies). It is also necessary to state the terms in the event of the mating being unsuccessful. This may offer a repeat mating at the bitch's next season, or a refund of the fee if no live or dead pups result from the mating.**
5. **a nice touch is to provide, in hard copy or digitally, photos of as many of the direct ancestors as you can. A pedigree is, otherwise, simply a list of names. Knowing what went into the mix will be useful for them when the puppies are born – and the future owners will like this as well.**

For my own interest I like to take a copy of the bitch's pedigree and also take a photo of her. That way I have some record of what she is like, which may be reflected in any of the pups she produces.

Following up

The wise stud dog owner will take an active interest in his dog's progeny. Ask the owner of the bitch for contact details of the purchasers, so that you can find out if they are being used for work or show. Try to get photos of them for your website, and also be prepared to offer help and advice if requested. The better your dog's produce performs the more he will be used at stud.

How long can a stud dog work for?

This will depend very much on the health and vigour of your dog. As long as he is interested in bitches, and is prepared to mate with them, he should continue to be capable of siring puppies. There is no reason to think his sperm will be any less fertile as he ages. If you have any doubts you can always get a sample of his semen collected by a specialist vet (your own local practice can refer you) and they can assess it under a microscope.

Artificial insemination

Some older dogs may suffer from prostate problems, which can impair their ability to mate bitches, reduce fertility and eventually lead to complete impotence. Working dogs can also sustain injuries to their vital equipment, which can damage them physically to the extent that natural matings may become impossible. Here again specialist veterinary companies can still, sometimes, obtain live semen from your dog, which can be inseminated into a bitch artificially.

Frozen Semen

In breeds with a small number of individuals you may get very infrequent opportunities to use your dog at stud. This will limit your future chances of using his offspring in any long-term breeding programme. It is worth considering the option of freezing a quantity of his semen, so that you can revisit his genetic legacy in the future. However, frozen canine semen is notoriously fragile, and it can be an expensive and unsuccessful experiment.

If you are lucky enough to own a stud dog of international interest, you may be asked to send frozen semen abroad. Do stress that fertility is hard to guarantee, and include some provision for this in your terms of stud service.

Infertility in a male dog

This condition is rarely met with, but can occur. The collection and examination of live semen *before* advertising your dog at stud (see **Artificial insemination** above) will enable you to campaign his services with confidence. For the owner of any potential stud dog the first mating is a traumatic time. The ideal first encounter, with an experienced brood bitch, is not always possible, and your dog may be faced with an aggressive or exceedingly nervous bitch. With patient handling all should be fine, but a bad introduction to his career could easily see it come to an abrupt and untimely end.

Even with a first successful mating under his belt, you are then faced with an anxious wait to see if the bitch is in whelp. If she is not, then you cannot be sure if either parent is infertile, if the bitch has had an infection which has caused her to re-absorb her pups or, horror of horrors – your dog is firing blanks! Only when he has pups 'on the ground' can you relax. Then you must visit the litter at about five or six weeks of age, to see if the pups are of a consistent type and representative of their sire. Only when you know that he is fertile, and stamps his stock with his own type (perhaps good heads or conformation), can you relax and campaign his services with some confidence.

If your dog sires puppies with any sort of physical problem, deformity or lack of vigour you **must** be prepared to subject him to examination to establish the root cause. This should include checking back through his pedigree and cross-questioning owners of any immediate ancestors, to find out if there is a history of such problems in his line. If there is you should stop using him at once. Sometimes owners refuse to admit that their 'boy' has any responsibility for shortcomings in his stock (another form of kennel blindness), and keep offering his services to bitch owners. What you must do, if this disaster happens to you, is to cease using him, notify the Breed Health Representative and make his breeding known. You should also consider castration.

The likelihood is that none of these disasters will happen to your dog. If all goes well, he will contribute to a new generation, and leave his breed better for it. Every breeder wants his or her own stock to achieve this lasting heritage. Good luck!

Index

Page numbers in bold font signify a primary entry for the topic